ROAD COMMUNICATION
IN
NORTH-EAST INDIA AND BEYOND

About the Editor

Tarun Dutta, Assistant Professor in History (Assam University Diphu Campus) has completed his Post-Graduation in History from Dibrugarh University and Awarded Master of Philosophy (M.Phil.) and Doctor of Philosophy (Ph.D.) by Assam University, Silchar, Assam. He has been awarded numerous fellowships including the Indian Council of Historical Research- Junior Research Fellowship. He has authored and edited many books and publications.

ROAD COMMUNICATION IN NORTH-EAST INDIA AND BEYOND

Bridging Culture and People

Edited by
Tarun Dutta

CONCEPT PUBLISHING COMPANY PVT. LTD.
NEW DELHI-110059

ISBN-13: 978-93-86682-37-6

First Published 2018

Published and Printed by

Concept Publishing Company Pvt. Ltd.
A/15-16, Commercial Block, Mohan Garden,
New Delhi-110059 (INDIA)
T : 91 11 25351460, 91 11 25351794
E : publishing@conceptpub.com; *W* : www.conceptpub.com

Cataloging in Publication Data--Courtesy: D.K. Agencies (P) Ltd. <docinfo@dkagencies.com>

Seminar on Road Communication and its Development in North-East India and Beyond : Bridging Culture and People (2016 : Diphu, India)

Road communication in North-East India and beyond : bridging culture and people / edited by Tarun Dutta.

pages cm

"The papers included in this volume were presented at a Seminar on Road Communication and its Development in North-East India and Beyond : Bridging Culture and People, organised by Department of History, Assam University Diphu Campus, on 29th to 30thSeptember 2016"—Acknowledgements.

Includes bibliographical references and index.

ISBN 9789386682376 (hardback)

1. Highway communications—India, Northeastern—History—19th century—Congresses. 2. Highway communications—Social aspects—India, Northeastern—Congresses. 3. Highway communications—Economic aspects—India, Northeastern—Congresses. 4. Economic development—India, Northeastern—Congresses. 5. Ethnic relations—India, Northeastern Congresses. I. Assam University (Silchar, India). Diphu Campus.Department of History, organizer. II. Title.

LCC TE228.3.S46 2016 DDC 388.31209541 23

Acknowledgements

The papers included in this volume were presented at a Seminar on *Road Communication and Its Development in North-East India and Beyond: Bridging Culture and People*, organised by Department of History, Assam University Diphu Campus, on 29th to 30th September 2016. The primary objective of the seminar was to discuss various issues of the communication problem in North-east India and its allied issues. We highly obliged to Prof. Sarah Hillaly, Professor in History, Rajiv Gandhi University, who gave the keynote address at the Seminar with an exposition of the basic issues of the road transportation of North-East India.

We remain very thankful to Indian Council of Social Science Research, New Delhi, Vice-Chancellor, Assam University and Pro-Vice Chancellor, AUDC for giving generous support to undertake this programme.

We wish to record our appreciation of the effort of Dr. Santosh Hasnu, Hans Raj College, Delhi, Prof. Kalyan Das, OKDISCD, Guwahati, Mr. Mrinal Gohain, Regional Manager, ActionAid Association, who gave their plenary/special lecture in the seminar.

We gave our special thanks to all organising committee members and well-wishers for their strong support and assistance towards the seminar.

To Mr. A.K. Mittal, Managing Director of Concept Publishing Co. Pvt. Ltd., New Delhi, we owe a debt of gratitude for agreeing to publish this volume expeditiously.

Tarun Dutta

Contents

Section B
Culture and Connectivity

Section C
Region and Beyond

Section D
Road and Its Impact

List of Contributors

Dr. Bijoy Mohan, Associate Professor, Department of History, Sibsagar College, Joysagar-785665, Assam.

Mr. Console Zamreinao Shimrei, Research Scholar, Department of History, Assam University, Diphu Campus, Karbi Anglong, Diphu-782460, Assam.
E-mail : *czshimrei@gmail.com*.

Dr. Dibyajyoti Dutta, Assistant Professor, Department of Political Science, Dibrugarh University, Dibrugarh, Assam.
E-mail : *dibyajyoti@dibru.ac.in*

Mr. Habib Fazlul Basid, Research Scholar, Department of History, Gauhati University, Gopinath Bordoi Nagar, Guwahati-781014, Assam.
E-mail: *habib. basid@gmail.com*

Ms. Himadri Gogoi, Research Scholar, Department of History, Assam University, Diphu Campus, Diphu-781014, Assam.
E-mail: *hiyax5774@gmail.com*

Mr. I. Imkongakum Ao, Research Scholar, Department of History, Assam University, Diphu Campus, Diphu-782460, Assam. E-mail: *mozerakum@gmail.com*

Dr. Jagadish Nath, Associate Professor of Marketing, School of Business, The Assam Kaziranga University, Koraikhowa, NH 37, Jorhat, Assam.
E-mail: *jagdish@kazirangauniversity.in*

Dr. Jitu Saikia, Assistant Professor, Department of Economics, G.C. College, Silchar, Assam.
E-mail : *saikiajitu 2008@gmail.com*

Ms. Juli Tokbipi, Research Scholar, Department of Economics, Savitrybai Phule Pune University, Ganeshkhind, Pune-411007, Maharashtra.
E-mail : *juliz_06@yahoo.com.*

Ms. Jushna Beypi, Research Scholar and RGNF, Department of Anthropology, Assam University, Diphu Campus, Karbi Anglong, Diphu-782460, Assam.
E-mail: *jushnadpu@rediffmail. com*

Mr. Kezhangulie Metha, Research Scholar, Department of History, Assam University, Diphu Campus, Karbi Anglong, Diphu-782460, Assam, India.
E-mail: *kezhametha@gmail.com*

Dr. Mayengbam Nandakishwor Singh, Assistant Professor of Political Science in National Law University and Judicial Academy, Guwahati Hajo Road, Amingaon, Guwahati-781031 Assam.
E-mail: *nkishwors@gmail.com*

Mr. Mintu Tokbipi, Research Scholar, Department of Economics, Savitrybai Phule Pune University, Ganeshkhind, Pune-411007, Maharashtra.
E-mail : *mtokbi007@gmail.com*

Dr. Mokhalles M. Mehdi, Assistant Professor of Marketing, MIS & IT, School of Business, The Assam Kaziranga University, Koraikhowa, NH 37, Jorhat, Assam.
E-mail: *mokhalles@gmail.com*

Mr. Narendra Kr. Das, Department of Geography, Digboi College, P.O. Digboi-786171, Assam.
E-mail: *narendrakumardas@gmail. com*

Dr. Paul B. Chonzik, Associate Professor, Department of History, Assam University, Diphu Campus, Karbi Anglong, Diphu-782460, Assam, India.
E-mail : *pchonzik@gmail.com*

Ms. Rengmini Tokbipi, Research Scholar, Department of Anthropology, Assam University, Diphu Campus, Karbi Anglong, Diphu-782460, Assam.
E-mail : *rengminit@gmail.com.*

Ms. Rosanna Kikon, Research Scholar, Department of History, Assam University, Diphu Campus, Diphu-782460, Assam.
E-mail: *rosakikon@gmail.com*

Mrs. Sangeeta Boruah Saikia, Department of Geography Digboi College, P.O. Digboi-786171, Assam.
E-mail: *saikiasangeeta@ yahoo.com*

Dr. Santosh Hasnu, Assistant Professor, Department of History, Hansraj College, University of Delhi, Delhi-110007.
E-mail: *rexhansu@gmail.com*

Prof. Sarah Hillaly, Professor, Department of History, Rajiv Gandhi University, Doimukh, Itanagar, Arunachal Pradesh.
E-mail: *sarahhillaly@gmail.com*

Dr. Subit Dutta, Assistant Professor, Department of Commerce, Assam University, Diphu Campus, Diphu-782460, Assam.
E-mail: *subit.dutta@gmail.com*

Ms. Sumitra Bithi Kachari, Phalengichuk Gaon, Titabor, Jorhat, Assam.
E-mail : *kacharibithi08@gmail.com*

Dr. Tahir Hussain Ansari, Assistant Professor, Department of History, Assam University, Diphu Campus, Karbi Anglong, Diphu-782460, Assam.
E-mail: *tansari786@gmail.com*

Dr. Tarun Dutta, Assistant Professor, Department of History, Assam University, Diphu Campus, Diphu-782460, Assam.
E-mail: *duttarun796@gmail.com*

Mr. Tarun Kr. Halder, Research Scholar, Department of History, Gopinath Bordoi Nagar, Guwahati University, Guwahati-781014, Assam.
E-mail : *tarunkrhalder@rediffmail.com*

Dr. Tridib Mahanta, Associate Professor, Department of History, Sibsagar College, Joysagar-785665, Assam.
E-mail: *tridimahanta@gmail.com*

Introduction

A road not only connects two places, it connects people to people, bears one culture to another. From very early period communications developed in various forms with different aims and objectives. In the passage of time, the necessities of these communications means were felt differently. Among those means of communication road communication or overland communication always occupies much importance than others. Hence, it received royal patronage at various times for its development.

In India, there were constructed several royal roads to serve the purpose of its construction. In North-east region of India also developed some such roads, which filled up the necessities of internal communication. Assam, from its historical evidence, it is presumed that there was a good network of land routes in ancient Assam apart from that other communication line were mentioned in various records. In the centuries before the advent of British in the region, there was a high degree of intermingling of various streams of people who came from East and West and entered in this region through these routes. Administration, trade and commerce were another important factors in the development of road during this period. With the outbreak of the Second World War, the transport difficulties became more acute and the region needed a secured road communication. Consequently, the construction of roads were given more priority as per the needs of defence.

North-east India is a frontier region situated in a geographically strategic location near the borders of the Indian sub-continent and has links with China, Bhutan, Nepal, Bangladesh and Myanmar. The region is culturally and racially diverse from the rest of India. The diversity of culture has been brought about by the intermingling of people from countries such as China, Bhutan, Myanmar etc. There have also been significant socio-economic and religious influences on the North-eastern region from its neighbouring countries, which were manifested through various means, such as land routes, maritime contact etc. The numerous passes in the mountains and hills of North-east India exhibit evidence that there existed strong socio-cultural and economic ties between North-east India and the neighbouring countries. Through these land routes, people of all kinds, ranging from travellers and invaders to traders and adventurers, entered the region. *"The frontier's numerous passes has received people from China and various parts of South East Asia, invasions took place and allowed trade relations to exist"*.[1]

Since the North-eastern region geopolitically shares a long frontier with China, Tibet, Bhutan, Bangladesh and Myanmar, this frontier region has experienced considerable population movement from different directions. The region has maintained a few trade and migration routes with East and South-east Asia. In the centuries before the advent of British in the region, there was a high degree of intermingling of various streams of people. People came from far away countries such as China, Mongolia, Thailand, Laos, Tibet and Burma over a long period. As Bridget and Raymond Allchin point out, *"This North-eastern frontier region is often regarded as a colourful corridor between South and South East Asia because it has been receiving various races, languages, cultures and religions from South East Asia and China making the region culturally and racially diverse"*.[2]

The frontier episode of North-east India is important for its strategic, socio-economic, military and cartographic position. With this great importance from the point of people, connectivity—both internal and external, cultural interaction, immigration and frontier war, the book is trying to give a look into the process of development of road connectivity and its relationship with people etc.

Tarun Dutta
Department of History
Assam University Diphu Campus
Assam

Notes

1. R.N. Reid, 1966, *Years of Change in Bengal and Assam*, London: Ernest Benn, p. 75.
2. Allchin, Bridget and Raymond, 1968, *The Birth of Indian Civilization*, London: Pelican, p. 23.

Section A
History, Politics and Development

1

Road Communication and Its Development in North-east India and Beyond within a Historical Framework

Sarah Hillaly

Introduction

The historiography of South Asia in the realm of transport has been built on the premise of how improved means of communication through the network of roads, railways and canals were 'public works', sought to improve the lives of the people. Lack of transport and communication in conjuncture with trade facilities has been viewed as being in a state of 'barbarism' within colonial historiography. Transport facilities for commercial purposes came to be viewed as a measure of triumph for western techno-science, fostering regional development on one hand, while unifying regions to the world trade cycles—particularly in the railways. It was linked with ideas of material and moral progress rendered by the efficient road communications. The debates on economic nationalism too rested on whether transport and communication of all kinds were beneficial to the British or the Indian nation.

These ideas failed to take into consideration as to how imperial techno-science interacted with specific local realities and emerged as the science of the empire. According to critiques like Sandra Harding the progress of European science is chiefly attributable to the "military economic and power of European cultures" rather than to claims of possessing greater rationality or commitment to the pursuit of disinterested truth.[1]

The unification of regions or fostering of regions created new socio-economic constellations within which the indigenous business interacted. Ravi Ahuja in his book *Pathways of the Empire: Circulation, 'Public Works' and Social Space in Colonial Orissa, c. 1780–1914* tries to argue that each of these infrastructure modes have been studied, de-contextualised from its historical contexts and considered to have no specific context. It was presumed that both roads and railways catered to a wide range of social actors and hence were all-pervasive in its impact. He further states that colonial sources were extremely laden with data on railways and canals, with a corresponding paucity of data on roadways and waterways. To him it reflected the dominance of certain social interests.[2]

Within the history of South Asian transport and infrastructure Jean Deloche's *Transport and Communication in India Prior to Steam Locomotion,* 2 Vols., (New Delhi: Oxford University Press,1993-1994) has been significant in tracing the pre-colonial modes of transport. Such modes of transport within the historiography have received very little attention, except in terms of alluding to roads as gleaned from the *Arthashastra* of pack bullocks wherein comparisons are made of its relative costs and contrasting it with savings achieved by the coming of railways.[3]

A recurring trope among historians of South Asia's transport infrastructure continues to be mediated by notions of pre-colonial self-sufficient villages in a permanent state of stasis. The argument stemming from it is that all means of communications—roads, railways and canals—was the handiwork of the colonial state. Jean Deloche "reconstructed spatial patterns and technological modes of transport in pre-colonial South Asia proving beyond doubt that road engineering and bridge building pre-dated British innovations."[4] Fernand Braudel too in his work on the Mediterranean had pointed out that roads were "no mere tapes unrolled over land." They were rather contingent on the existence on modes of economic, political and cultural institutions, which included "ports and towns, fairs and markets, pilgrimages and festivals, caravan serais and mail services, toll stations and police posts".[5]

Within the historical context, development of transport infrastructure is essentially entwined with the socio-economic, political and cultural processes of the corresponding period. Ravi Ahuja alludes to road-building measures by the 'little kingdoms' of the early nineteenth century was directed to serve political and cultural needs of the polities, rather than to seeking to centralise control. The lack of a vibrant road network was therefore not to be construed as 'neglect' of infrastructure by the earlier regimes. Tarashankar Banerjee while reflecting on the importance of roads remarks that, "roads are a vital factor in the national economy. They link industry with agriculture, producer with consumers, and the external market with the internal one."

The Context

In the context of colonial Assam my book on *The Railways in Assam 1885-1947* (Varanasi: Pilgrims Publishing, 2007) which emerged out of my Ph.D. dissertation, is a maiden history of transport history. It was framed within the discourse of transport and communication, where the railways visualised within the older frame of discourse as a harbinger of major technological changes. The commonalities and the distinctiveness of the railway policy in terms of its introduction is dealt with. As a 'public work' the role of railways in all sectors of the economy, urbanisation and immigration as a concomitant has been explored. The social effects too have been explored in the work. A limited critique of the colonial policies has been attempted to account for the lopsided development concomitant to the growth of an "enclave economy" centred around the plantation, petroleum and coal mining industries. The broader framework was to view railways as a "productive public works". Beyond the purely economic implications the study has not taken into account its core social and cultural implications.

One of the chapters however has dealt with the pre-railway communication and transport, wherein a sketch of the pre-colonial roads and waterways has been made. It also discusses the roadways constructed by the British and the limited expansion of steam shipping. Their role not only in the "opening-up of the country" but to meet "strategic needs" have been highlighted. Despite an attempt to draw continuities from the pre-colonial modes of transport, the circulatory regime and its social context has not been investigated.

Roadways in North-east India

The pre-colonial period in Assam under the Ahom and Koch rulers witnessed the construction of large sections of embankment-*cum*-roads, which sustained internal communication within the Province. This network of roads impresses John M'Cosh an administrator ethnographer in his early travel writings. He writes thus :

> Very few provinces in India have been provided with such a splendid system of public roads as in Assam and from the great highways, which were carried uninterruptedly through the whole country to the great cross-roads, between the principal towns and their minute ramification, which connected the villages.[6]

M'Cosh alludes to Assam's earliest commercial relations with the outside world sustained through the Brahmaputra River. In addition, he refers to the existence of three overland routes to Bengal:

> The first was *via* Murshidabad, Malda, Dinajpore, Rangpore, (*sic*) Bugwah and Goalpara ... the second road was *via* Dacca, Dumary, Pacualoes (*sic*), Jumalpore, Singimari and Goalpara. The third *via* Sylhet, Cherra, Mophlung, Nungkhlow, Ranigodown, Canneeymookh and Guwahati ...[7]

These imperial road-*cum*-embankments had fallen into disarray, mostly being over-grown with jungles, which rendered them useless. This infrastructure had been built and maintained through collective efforts of a huge militia of freeman or *paiks*. Such *paiks* were under obligation to offer labour-rent to the state as a workforce in agriculture, and undertaking large-scale public works like building roads and tanks. The most significant ones constructed in the medieval period were the *Gosain Kamal Ali, Dhodar Ali, Dhai Ali* and the *Bar-Baruah Ali*.[8]

Modes of exchange with the overland neighbours, particularly the hills in the north and east were sustained through this network. Both inter- and intra-regional trade was clearly discernible through the networks, though its volume was small and constrained by the socio-political and economic structures of the medieval regimes. Each road network marked the convergence of social interests rather than purely commercial ones. Such networks were sustained through fragile 'pathways' which mediated flows of humans, goods and ideas.

The colonial intervention from 1826 inaugurated newer modes of economic control. The initial years were focused on maximizing revenues on land through commercialization of agriculture and subsequent monetization of the economy. Captain Jenkins, the Agent to the Governor General visualized expansion of trade in raw-produce as a concomitant to agricultural expansion. An imperative to such an expansion could only happen with a consequent improvement of the existing communication networks by land and water. In his letter to A.J.M. Mills in 1853, he emphasized on the need of improved communications thus:

> The measure for accelerating the progress of the Province.... A systematic improvement of the main roads is one of the obvious means we have.... A great obstacle now to the clearing of wastes is the want of communication between these tracts and the cultivated districts.... The outlay on the roads, if the works be regularly conducted and permanently established and be not again allowed to fall to disrepair, will as certainly be returned to Government by increase in revenue.... The people cannot [live] amongst wastes that are cut off from all communication with the present villages, but wherever the jungles are opened out, colonization does and will take place.[9]

In the colonial schema, the presence of good roads was linked to notions of permanence, civility and superiority of civilization.[10] Hence, communication becomes a site and a medium of grappling with the shifting economic structures; to understand trade and travel read with ideologies and discourses of colonial rule.

Responsibility of building roads and maintenance in India was vested with the Military road in each Presidency until 1854, The roads constructed though patchy with the exception of a few arterial ones were because of military requirements. In 1854, Lord

Dalhousie constituted a separate Public Works Department.[11] As 'public works' was no longer under the Military Board, increasing attention was paid to 'civil works' particularly civil buildings, roads, establishments, canals and irrigation works. In the preceding years, funds available for public works was a paltry Rs. 60 lakhs in 1849-50, which rose to nearly Rs. 2.25 crores in 1856-57.[12] Therefore, the East India Company which had totally neglected 'public works' for the greater duration of its rule showed a spate of activity in road building during the last years of its rule.

During the earlier colonial period, focus on road constructed was to enable speedy movement of troops, in the eventuality of renewed Burmese attacks and incursions from frontier tribes. Here too public works were vested with the Military Board. The initial years were spent on restoring "portions of the great roads of Upper Assam."[13] As Assam continued to be administered from Bengal in 1835 an Executive Officer operating from Dacca was appointed to oversee the communication network. Faced with delays and difficulties in execution Captain Jenkins demanded for employing an European overseer in Assam to ease the matter. Finally, in 1840 a Division of Public Works was created with Executive Engineer appointed for Upper Assam and an Executive Officer for Lower Assam, both subordinated to the Chief Engineer in Bengal.[14]

These military roads remained unmetalled, in most cases, cleared, and repaired, just enough for the passage of troops and elephants. These roads, because of their temporary nature, had to be constantly repaired. As the political situation gradually stabilized, roads for purely military purposes were rarely constructed. Construction and main tenance of all *kutcha* roads were left to the local authorities considered better equipped with the local milieu hence could provide local assistance in the form of land grants, labour and revenue. This was because the demands on the "Imperial revenue" was substantially heavy. The Government encouraged raising of funds through public subscriptions too Military roads continued to be managed from ferry funds raised in each district.

The Government encouraged raising of funds public subscriptions too. Public works undertaken by public money were not thoroughly executed by the Government officers. The public too, after having paid the money, showed no interest in its execution. In Upper Assam, many old and minor roads were cleared of jungles and repaired with the help of tea planters in the immediate vicinity of their gardens.

Paucity of funds remained a constant hindrance for local improvements. In order to tap alternate sources of funds, in 1861 Henry Hopkinson, the Commissioner of Assam, proposed to double land tax, so as to devote the excess so obtained, for works of public utility. The Board of Revenue did not consider it expedient to double the land tax.[15]

The Revenue Board extended the provisions of the Government Order No. 1847, of 28th July, 1859, sanctioning "the appropriation of 3 per cent on the collections to purposes of local improvements."[16] In 1862, the Government permitted the use of "the surplus proceeds of 1 per cent on the Income Tax Collection."[17] The creation of a full-fledged Public Works Department for Assam in 1868 ensured financial stability and streamlined its working. Further, by the Act X of 1877 District Road Cess was levied on all immovable property, within the district for raising funds for construction and maintenance of roads. It also provided for the constitution of 'Local District Committees' for the assessment and management of the "Road Cess."[18] The funds for the Trunk roads and Military Roads were drawn from the "Imperial Funds."

In the early years of British rule road alignments were decided on their military feasibility. These military roads were to provide render speedy troop movement, in the eventuality of renewed Burmese attacks or incursions of the frontier tribes. The initial focus was therefore on restoring "portions of the great roads of Upper Assam."[19] In 1840, alarmed by reports of Burmese attempts to open a road through the Naga Hills from the Kyendween (*sic*) river, made it imperative to improve the line of communication between the Muttock country and the troops reserve at Sibsagar.[20] Repairs were undertaken on the *Bar Baruah Ali*, the high road from Sibsagar towards Dibrugarh through Chokey Hat on the Buri Dehing. Portion of the high roads between Jaipore and Sibsagar was opened. This road was to be:

> ...of great political importance, as affording the means of moving troops and of controlling the Nagas as well as highly valuable as one of the roads of internal communication on a line, by which much traffic in salt is conveyed.[21]

The other roads repaired were from Dibrugarh to Saikwah and Jaipore to "facilitate movement of the Assam Light Infantry which has (*sic*) to guard the extensive frontier."[22] The *Bund* road from Sibsagar to Dikhu-mookh, according to Jenkins was of "highest utility in a military and political view."[23] Captain Butler, the Principal Assistant of Nowgong in a letter to Jenkins urged on the opening of two separate roads along the frontier of the district in view of the operations against the Nagas. The first alignment was to be from Nowgong *via* Samuguting to Dimapur, the other from Golaghat to Dimapur.[24] To complete the line of communication along the frontier posts, a road was constructed between Dum-duma and Makoom.[25] A small stretch of road between Golaghat and Nagorah was opened, as it was "much frequented by the Angami Nagas and our traders and troops."[26] These military roads were unmetalled, in most cases, cleared, and repaired, just enough to enable the passage of troops and elephants.

The Government too put in its effort to open branch roads from the main line of the tea gardens to the landing places (*ghats*) on the Brahmaputra. In the district of Sibsagar, portions of the *Dhai Ali* between Sibsagar to Buri Dihing Mookh; *Barpatra Gohain Ali* from Sibsagar to Disang, and the *Sologuri Ali* from Disang to Buri-Dehing were constructed.[27]

The Government emphasized on the need for road building into the hills, to establish markets and develop them, as "roads and markets ought very speedily to create a social revolution in the hills."[28] A road was constructed between Udalguri in the Kureaparah Doar and Mangaldoi on the banks of the Brahmaputra, with the object of "inducing the Bhutias to come down to Mangaldoi (*sic*) for the annual fair."[29] A road was opened from Goalpara to the eastern *hats* of Jirah, Damrah and Nibari. The road was of much importance to the merchants of Goalpara as there was a large trade in cotton.[30]

The principal roads in the well-populated parts of Kamroop were repaired. Repair of the great western road, connecting Guwahati to Goalpara and therefrom to Bengal was taken-up. The *Gosain Kamal Ali* was restored between Manas and Bur-Nadi. The high *bund* road extending from Guwahati through East Dimarua to Raha, across the Kullung along its right bank crossing Nowgong, continuing up to Jorhat and Sibsagar was restored.[31]

The tea industry around which all the developmental activities in Assam centred, played a crucial role in the growth and development of roadways:

> As early as 1845, the Assam Company claimed with some exaggeration, that it had opened or repaired some 1,280 kilometres of public road, had erected 266 bridges and established several ferries across the rivers.[32]

The roads mostly linked the gardens to landing places (*ghation*) on the Brahmaputra from where country boats shipped them to Calcutta. Roads linked the gardens with their neighbouring villages, which enabled cultivators increased facilities for disposing of the produce of their lands.

The first road, conceived for external communication connecting Guwahati with Sylhet was considered crucial for troop movements in case of fresh Burmese incursions. As early as 1826, David Scott persuaded the Khasi Chiefs of Khyrim and Mylliem to agree on the construction of a road through their territory.[33] Work commenced on the route in 1829, but had to be stalled because of the murder of Lt. Burlton and Beddingfield, who were in charge of the construction. Following the annexation of Khasi Hills in 1832, interest was renewed on the project. However, the military imperative changed after 1840, when Jenkins insisted on the road aligned through the Khasi Hills was: "the road was of highest importance to the bordering countries in a commercial point of view…also equally good with any other if not the best for military operations of any unforeseen exigencies".[34]

Colonel Hopkinson in 1862, in a letter to the District Officers on the issue of opening-up of trunk lines through the province favoured an early construction of the Guwahati -Sylhet road, as he perceived that:

> communication with Bengal will be sooner established by it, because with the railway to Dacca, it will probably be the quickest route of the Presidency, as for mails to reach as from Calcutta, because it at once renders accessible large tract of country perfectly suited to European colonization and because of the community of interest between Cachar and Assam which a road also uniting the two provinces together will so well serve.[35]

The alignment was finally decided upon and sanctioned in 1863. The route selected was between Guwahati in Assam *via* Shillong to Chattak on the River Soormah (*sic*), a distance of 104 miles. In 1866, a grand project for a 'Trunk Road' running through all the districts of Assam was conceived. The project was formulated on a much larger scale than funds could permit. After assessing the financial implications of the project and based on the availability of funds, the idea of a "railway formation for the Trunk Road was given-up and orders were issued to continue work on a smaller scale of a sixteen feet top width." [36] Carrying regular mails, toll stations constituted as one of the important practices.

Till the end of the nineteenth century, the reports of the Assam Tea Association featured complaints on status of the existing roads, where the proportion of the metalled roads to unmetalled was insignificant. Since 1896 cart traffic played a significant role with Manipur and its neighbouring areas. Despite bad communication, the trade from the uplands north of Assam and the Lushai and Naga Hills including Hill Tipperah approximated at about Rs. 13 lakhs.[37]

In 1911, the Indian Tea Association stressed on priority being acceded to development of roadways. With the revocation of the first partition of Bengal in 1912 emphasis was laid on development of roads wherein Rs. 41 lakhs as special grants for roads were released. The first passenger service was inaugurated between Guwahati and Shillong in 1916. However, its reach was limited to goods carried for the Planters' Stores Limited. The condition of the aforementioned road led to a poor response by a Calcutta firm to run a passenger service.[38] Finally, in 1922, the first passenger bus service was started between Jorhat and Jhanji in Upper Assam. Both mail and passengers to Shillong were carried by *Tongas* till the introduction of mail service in 1922. Close to 400-600 bullock carts, with a carrying capacity of 15-20 maunds carried provisions like salt, cloth and other essential services, in return bringing back potatoes and oranges. [39]

During the Second World War, road construction received priority to meet the defence requirements. By the end of 1942, an acute shortage was witnessed as all vehicles were requisitioned by the Assam Transport Organisation. A private enterprise under European management stepped in to increase its fleet to 400 vehicles and employed 1,100 people within a span of six months.

During the entire duration of the war about 90 per cent of its fleet was engaged in providing logistical support to the military, while the rest catered to civil supplies. As the war ended Assam Transport Organisation went on the verge of liquidation. Government decided to intervene and on the recommendations of a Committee they renamed it as State Transport Corporation in 1948.[40]

The history of the Stilwell road as a part of the war effort against Japan, to maintain American-British link with China is one of the significant infrastructures created during the war.

The Trajectory of Transport History

With the spatial turn in historical debates, the idea of production of space of Henri Lefebvre is being used for analysing issues of circulation and transport. The rise of railways and steamshipping as modes of transport in the 19th century not only ushered in a revolution, but contributed crucially to transformation in social space.[41] The conception of annihilation of space and time was evident in the early discourses on road building across India where the entire country was viewed as a blank slate with roadless plains and impassable rivers. It was colonial imperative to open-up these spaces to civilization, where the new recreation of spatial environment in the form of roads were planned, thus making claims to creating new modes of circulation. The pre-colonial modes of transport found little recognition as they were always mentioned in the passing.

Currently, the term communication has acquired a disembodied character with the waves. Therefore, the construction and utilization of roads, waterways and railways become spatial practice of 'transport'. The ideas promoted were that new forms of transportation in concrete historical terms brought in social domination by privileging ideas of spaces, fortifying dominant interests, which subordinated interests were undermined. Such ideas fails to encapsulate the contradictions inherent in terms of the multilayered historical inheritance and polyvalence of the built environments. An effective study should be able to not only look at polyvalence at level of the nation, but in regional and local settings and also include the socio-economic and cultural fieids wherein all daily experiences are wherein such transport infrastructure was used very sparingly.

Arguments like infrastructure helping to link all local networks into the world trade cycles can be re-examined in terms of looking at the process of fission/departure by micro-level studies. All practices of space is defined by rhythmic cycles of nature specific to each area. Periods of circulation were defined by seasons, largely agricultural/non-agricultural, religious gatherings, fairs and festivals. Even public works like building roads, ponds, bridges etc. tried to follow the seasonal cycles though at times such norms were deviated from. The creation of new transport infrastructure which was characterized by the productive and reproductive cycles of capital imposed its own cycles over that of nature. One can therefore study circulatory flows of goods, ideas and human movements not to understand the linearity of the flows, but the negotiations with newer modes of transport and creation of mediated flows.

Notes

1. Celia Roberts and Adrian Mackenzie, 'Science' in *Theory, Culture & Society*, Vol. 23(2–3), 2006, p. 158.
2. Ravi Ahuja, 2009, *Pathways of the Empire: Circulation, 'Public Works' and Social Space in Colonial Orissa, c. 1780–1914*, New Delhi: Orient Black Swan, p. 3.
3. Ravi Ahuja, *Pathways of the Empire*, p. 4.
4. Ravi Ahuja, 2004, "Opening up the Country? Patterns of Circulation and Politics of Communication in Early Colonial Orissa", *Studies in History*, 20:1, p. 73.
5. *Ibid.*, p. 74.
6. The brother of the great Koch King Nara-Narayan, after whom the road was named, constructed the *Gosain Kamal Ali*. It extended from Sadiya to Kooch Behar, a distance of 350 miles. The *Bangal Ali* stretched from North-East Darrang to North Guwahati . The *Dhodar Ali* was in the southern portion of the Sibsagar district and the *Dhai Ali* was from Rangpur to Jaipore.
7. J. M'Cosh, *op. cit.*, pp. 8-9.
8. Sabyasachi Bhattacharjee, *Financial Foundations of the British Raj* (Shimla, 1971), p. 114.
9. A.J.M. Mills, *op. cit.*, Appendix 'B', No. 275 of 1853, from Colonel Jenkins to A.J. Moffat Mills, Para 2.
10. Nitin Sinha, 2012, *Communication and Colonialism in Eastern India: Bihar 1760s—1880s*, London: Anthem Press, p. xvii.

11. *Ibid.*
12. *Ibid.*
13. Assam State Archives (ASA), *Letters Issued to Government* (LIG), Volume No. 9, January-September 1840, No. 23, 28th January, 1840.
14. *BJC*, (Criminal), 1838, August 7th, No. 106.
15. B.G. Papers, 1861, File No. 388, Nos. 1-6, No. 494, dated 6th July, 1861.
16. *LIG, op. cit.*
17. *LRG*, Volume No. 44 (d), January 1859-1864, No. 931, 3rd March, 1862.
18. B.G. Papers, File No. 42/50 of 1873.
19. *LIG*, Volume No. 9, January-September 1840, No. 23, 28th January, 1840.
20. *LIG*, Volume No. 9, January-September 1840, No. 85, 4th May, 1840.
21. *LIG*, Volume No. 9, January-September 1840, No. 147, 28th August, 1840.
22. *LIG*, Volume No. 10, No. 131, 9th March, 1852.
23. *BJC*, (Criminal), No. 40, 19th September, 1842.
24. *LIG*, Volume No. 15, No. 112, 21st September, 1850.
25. *LIG*, Volume No. 15, No. 152, 28th November, 1850.
26. *Ibid.*
27. A.J.M. Mills, *op. cit.*, Appendix F-1-23, "List of Principle Roads in Sibsagar District".
28. B.G. Papers, Nos.1-3, File No. 15/18 (1873), *cf.* S.B. Medhi, *Transport System and Economic Development in Assam*, (Guwahati, 1975), p. 42.
29. *LIG*, Volume No. 16, No. 320, 10th August, 1852.
30. Goalpara Papers 1873, File No. 981, Serial No. 122, 28th February, 1873.
31. *BJC, (Criminal)*, No. 4, 13th October, 1835.
32. Amalendu Guha, MECA, p. 160.
33. For details of David Scotts' policy in the Khasi Hills, refer to Nirode K. Barooah, *op. cit.*
34. Ravi Ahuja, *op. cit.*, p. 27.
35. Letters issued to District Officers, Volume No. 9 (a), 1861-1864, No. 32, 31st January, 1862.
36. W.W. Hunter, *op. cit.*, p. 55.
37. S.B. Medhi, *Transport System and Economic Development in Assam*, Guwahati· Publication Board, 1978, p. 42.
38. *Ibid.*
39. *Ibid.*
40. *Ibid.*
41. Ravi Ahuja, *op. cit.*, p. 27.

2

Connectivity with East *via* North-east India: A Study

Tarun Dutta

Introduction

North-east India is a frontier region situated in a geographically strategic location near the borders of the Indian sub-continent and has links with China, Bhutan, Nepal, Bangladesh and Myanmar. The region is culturally and racially diverse from the rest of India. The diversity of culture has been brought about by the intermingling of people from countries such as China, Bhutan and Myanmar. There have also been significant socio-economic and religious influences on the North-eastern region from its neighbouring countries, which were manifested through various means, such as land routes, maritime contact etc. The numerous passes in the mountains and hills of North-east India exhibit evidence that there existed strong socio-cultural and economic ties between North-east India and the neighbouring countries. Through these land routes, people of all kinds, ranging from travellers and invaders to traders and adventurers, entered the region. "The frontier's numerous passes has received people from China and various parts of South-East Asia, invasions took place and allowed trade relations to exist."[1]

Since the North-eastern region geopolitically shares a long frontier with China, Tibet, Bhutan and Myanmar, this frontier region has experienced considerable population movement from different directions. The region has maintained a few trade and migration routes with East and South-east Asia. Centuries before the advent of British in the region, there was a high degree of intermingling of various streams of people. People came from far away countries such as China, Mongolia, Thailand, Laos, Tibet and Burma over a long period. As Bridget and Raymond Allchin point out, "This North-eastern frontier region is often regarded as a colourful corridor between South and South-East Asia because it has been receiving various races, languages, cultures and religions from South-East Asia and China making the region culturally and racially diverse."[2]

North-east India is a zone where..."South East Asia begins or ends, and it constitutes a transition zone where the Indo-Caucasian and Mongoloid racial stocks intermingle."[3] Most of the people of North-east India have their roots outside India and few groups have migrated to the region from different parts of the Indian heartland. For example, most of the tribes and castes of Assam and other North-eastern regions claim their origin to South East Asia, particularly Tibet, China and Myanmar and come through various routes. Once this migratory routes became important for its strategic, socio-economic, military and cartographic position. Among these routes Indo-China connecting routes very much importance, which later on developed to a military highway.

Development of Roads in Early Period

From the historical evidence, it has been proved that there was a good network of land routes in Ancient Assam. During the period of Bhaskarvarman, in the middle of the seventh century AD, the Chinese explorer Hiuen Tsang entered Magadha through Champa [Modern Bhagalpur, Kahangala (Rajmahal)] and Pundra Vardhan (Rangpur) moving eastward reached Kamrupa.[4] This signifies the existence of land routes from Magadha to Assam. From the account of Hiuen Tsang, it is found that China had a direct link with Assam through a mountainous land route. This route was from the upper Brahmaputra Valley to Burma and then to China. By this route, a considerable trade between China and Burma was carried on. The line of trade was from Sadiya, the frontier station of the Upper Brahmaputra Valley, *via* Bisa across the Patkai ranges of mountains and through the valley of Hukoong to the town of Mukung (situated on a navigable branch of the river Irrawaddy). The merchants proceeding from Mukung to Ava at once descended the Irrawaddy to the capital of Burma.[5] This route helped the Burmese to invade Assam. It is very hard to go outside from Assam before the advent of the colonial people. Strict security was imposed in all possible passes of communication, and without prior permission from the throne, no one was allowed to enter or go outside. Its earliest commercial relation was with its nearest province Bengal and main means of communication was River. From tributaries of Brahmaputra, the goods were collected and through it, the mercantile boats ply towards Bengal. John M'Cosh traced the existence of the route from Sadiya to Tibet across the Himalayas along the course of the river Brahmaputra.[6] Further, M'Cosh has mentioned the three inter-state (between Bengal and Assam) land routes. The first one through Mourshidabad-Malda-Dinajpur-Rangpur-Bugwoh and Goalpara in the western part of the Brahmaputra Valley; the second route was *via* Decca-Jamalpur-Singimara and Goalpara; the third route was through Sylhet, Chirra, Morphong, Nunglow, Ranny-Godown and Guwahati . However, all these routes were impassable in the rainy season and the traffic to Bengal was confined to water routes only. The important routes to Bhutan were Bijni-Duar, Buxa-Duar, Na-Duar, and Char-Duar. M'Cosh also noted that there were not less than five routes leading from Sadiya to China proper.[7] They were the passes of Dihing, the Mishmi route; the Phugan pass to Manchee and China; the route through Manipur to the Irrawaddy and the Patkai pass to Bhamo on the Irrawaddy.

Development of Roads in the Medieval Period

"Very few provinces in India have been provided with such a splendid system of public roads as in Assam and from the great highways, which were carried uninterruptedly through the whole country to the great crossroad, between the principal towns and their minutes ramification, which connected the villages".[8] During the Ahom rule, many important roads were constructed for defence purposes and maintained an excellent internal communication within the state. During the Muslim expedition to the Ahom territory in the middle of the sixteenth century, the Ahom soldiers constructed a road from Guwahati to the bank of the River Korotoya.[9] Again, during the war between the Konches and the Ahoms, the Konches constructed

an embanked 560 km long road from the capital of Cooch Behar to Nara-Narayanpur in the southernmost part of present North Lakhimpur sub-division. The road was named as 'Gohain Kamal Ali'. This great undertaking was completed in 1547.[10] During the Ahom rule, other roads such as the 'Bengal Ali' from north east of Darr (present Darrang) to North Guwahati ; 'Jaintia Road' from Rohachouki to Jaintiapur; 'Bar-Ali' from Jorhat to Rangpur; 'Dhai-Ali' from Rangpur to Jaipur and 'Dhodar-Ali' in the south of the Sibsagar district, *Bar-Baruah Ali, Borpatra Gohain Ali, Sologuri Ali* etc. were constructed by the *Paiks*.[11] Thus in the sixteenth and seventeenth centuries, a number of good roads were constructed for inter-and intra-regional communication. All the roads were fair weather and carts, horses and elephants were the chief carriers.

The Ahom rule, perhaps for their grand design for the expansion of their kingdom and for strategic reasons, laid considerable emphasis in roadway communication. This fact is evident from the above-noted facts, but owing to the absence of systematic records, it is difficult to present a clear picture of road communication planning and development during the Ahom rule over the Brahmaputra Valley region.

After the fall of the Ahom power, those roads became useless due to the lack of proper maintenance. In 1817, the Burmese invaders entered Assam through the Patkai route and used the old jungle infested roads for their movements. The political disturbances caused by the Burmese insurgents completely paralysed the normal lives in the Ahom kingdom.

Development of Roads in the Colonial Period

The absorption of Assam into the British Indian Territory was achieved in a phased manner. Lower Assam, with substantial revenue, was politically annexed in 1826 and Upper Assam was resumed back to old monarchy. The fateful discovery of tea was instrumental in the final annexation of Upper Assam in 1838. Wasteland grants were awarded the thrust on inducing peasants to cultivate cash crops. The barter-oriented economy was transformed into a monetised economy, and efforts were made to expand trade and commerce.[12]

After 1826, the armed force of East India Company had to face the risk of the life in the movement in the interior parts of

Assam. Later, the new ruler had to take steps for repairing and reopening the old routes for spreading the administrative control over the entire region. The roads which were constructed by King of Assam in sufficient width and with a little cost the road can be operational for wheeled traffic. The embankment can also be used for traffic with minor repair.[13]

A comprehensive scheme was downed up to repair and reconnect several roads in upper Assam. The Bengal Government was also involved in the matter, particularly, for the renovation of the strategically important roads in Assam.[14] To set an economic link with the Assam agricultural export-oriented market, the colonialist sought to maximise their benefits. It was realised that without a good communication by both land and water, all-round development of the province will be hampered. "The measure for accelerating the progress of the province....A systematic improvement of the main roads is one of the obvious means we have...A great obstacle now to the clearing of wastes in the want of communication between these tracts and the cultivated districts...An outlay on the roads, if the works are regulated conducted and permanently established and be not again allowed to fall into disrepair, will as certainly be returned to Government by the increase in revenue...The people cannot live amongst wastes that are cut-off from all communication with the present villages, but wherever the jungle are opened out, colonisation does and will take place."[15]

Linkage with North-east and China *via* Burma during the Colonial Time

The records of British's and their expeditions give us some details of the existence of road to China through Burma. John M'Cosh, a 19th century British officer in Bengal Medical Service published two major works '*Topography of Assam*' and '*On the various lines of overland communication between India and China*' published in the proceeding of the Geographical Society of London in 1861, mentioned five routes.

The first route was across the pass of the Dihong. The Dihong is said to be the "main tributary of the Brahmaputra", the termination of the great Tibetan River Tsan-Pu. Pilgrims passing by this route reach Mah-Lou, a frontier town of Tibet, in sixteen days. Four more days journey away is a popular city with a regular Chinese Government, called Rho-Shee-Mah. The Dihong pass is said to be always difficult and impracticable except in summer.

The second route is called the Mishmee route. It leads for a couple of days up to the Lohit to where the river ceases to be navigable and "thence by a footpath ten days further along its banks to a place very sacred in Hindu Mythology, called Brahmakund. Great numbers of Hindu pilgrims go there for absolution."

The third route is by the Phugans pass to Manchee and China. This is said to lead to the Wang-Leo-Bum mountain range at an altitude of 8400 feet, through a dense jungle of oak, pine, rhododendron and juniper where no footpath can be seen and where two British travellers, Wilcox and Burlters were led by their guides following the notches in tress, which they made in an earlier journey. This route leads to a point in the upper reaches of the Irrawaddy, where the river is fordable.

The fourth route is by the Patkai pass to Bhamo and then to China. This is the route which was followed by immigrant groups like Ahoms and invading Burmese in 1817-22.

The fifth route is straight to Phung and joins the Old Burma routes at Noika on N'Bento and goes on thence to the Hukoong valley.

Some other British officers, who made a survey in to connect with China through North-east were Captain Neufville, Lt. Charlton, Captain B.F. Hannay, R.B. Pemberton etc.

J.F. Needham, whom earlier mentioned, Assistant Political Officer at Sadiya, while at Calcutta received a telegraph from the Secretary to the Chief Commissioner, directed him to proceed to Shillong immediately, as the Chief Commissioner wished to discuss with him a proposal about leading a small column from Assam to Mungkham to meet a column which was proceeding towards Assam from Burma.[16]

In this connection Needham surveyed the route to Mungkham and reported that, "I feel sure from the enquiries which I made at Hukoong that this is the route which we should use on all future occasions on which we may send a force across the hills. The new route to be selected (from Assam) need not necessarily be *via* the Nongyong Lake and Namyung. We might still march straight to Phung, and join into the old Burma route at Noika on N'Bento and go on thence to the Hukoong valley".[17] Further, in the winter of 1917-18, Mr. Stevenson, the executive Engineer of the Assam Bengal Railways explored the Pataki ridge to a distance of 20 miles west at the ways pass and 25 miles east thereat and concluded that the lowest point in this ridge cash in the area surveyed by him was the Sympana Saddle. By 1919, the Government of India had done a preliminary survey of the Hukoong valley route. Once again in the year 1920-21, the British Government appointed F.W. Allen, C.B.E. as Engineer in-charge to take-up the survey of another route connecting Assam and Burma from Ledo to Sahnaw respectively. In his report submitted on April 30, 1922, Allen suggested three probable routes.

All these reports of missions done by British officers were come to used when the Second World War broke out and Japan closed the all contact with China by overland and maritime. In that situation, the Allied power decided to construct a road on basis of earlier surveys done by colonial people. In the same time to save China from Japanese occupation and Allies prestige, the road was constructed from Ledo in Assam, India to Kunming at Yunan province of China, over the Patkai hills. The work started on 16 December, 1942, in command of General Joseph Warren Stilwell. In this construction work local tribes men, tea garden labourers, Gorkha, Chinese etc. local people was engaged and engineers were also imported from America. Indian Tea Association provided U.S. engineers with 8,000 labourers for general road construction work.[18] The Tea Industry, at the request of the competent military authorities, also had undertaken the task of securing suitable labour for work on the Eastern Frontier Communications; the number required being 25,000 for the Manipur project and 75,000 for the Ledo project. The labourers engaged in the Ledo road work were basically tribal from these areas numbering 1200.[19] Simultaneously, 10,000 Chinese and Kachins were working from the other side starting from the Yunan-Burma Railway construction site. Two pre-fabricated Hamilton iron bridge were built over each stream and river.[20] 75,000 labourers well equipped with spades, axes and knives were supplied by the tea garden for the construction of the Ledo Road.[21] The majority of local workers were adept in the use of the *khodalie* or hoe, and the *dah,* a knife with a long straight blade. In addition to daily wages, which varied from 11 *annas* for the Pioneers to Rs. 2/4 for the ITA groups, they were given rations and quarters with the exception of the muster-roll labour. Although their labour accomplished a useful purpose, it was never wholly satisfactory. Most of the men were untrained, and the term of their contract ran only from three to six months. Excerpts from diaries of the soldiers speak of women recruits as labourers and the nature and conditions of the employment of the 'Indian coolies,' "they carried dirt and rock on their heads, there were tons of dirt to be moved....and those Indian coolies were a big help too...it was a little hard to get used to them at first...it was odd seeing women doing man's work....hard, physical labour...and *"they carried dirt and rock on their heads. Indian coolies were a big help too. It was a little hard to get used to them at first. It was odd seeing women doing man's work-hard, physical labour and they carried everything on their heads. The loads were tremendous—three or four coolies would be required to place a load on the head of another one."*[22] The materials were supplied from America and other parts of the colonies as lease land agreement. The construction process was stated in war footing and completed 1079 miles within two years under the supervision of General Joseph Warren Stilwell. On 12 January, 1945, General Pick had led the first convoy out of Ledo, bound for Kunming, China. The 113 vehicles, driven by representatives of all the engineer units that had worked on the road, consisted of heavy cargo trucks, jeeps, and ambulances. Among the passengers were some 65 radio, magazine, and newspaper correspondents. The convoy reached Myitkyina on 15 January, where it stayed until 23 January because of the tactical situation. On 28 January, Pick led the convoy into Wanting where T.V. Soong, the Chinese Minister of Foreign Affairs, welcomed him. On 4 February, the convoy reached Kunming as firecrackers exploded, missionary nuns waved, and Chinese bands played. When the Road was completed in January 1945, it had absorbed a huge amount of manpower and material resources. The price of its construction was very high, its total cost to the Allies being about $148,910,000. Approximately 17,000 American engineers had worked on the road and 1133 Americans had been killed. Thus in this chapter discuss the historical importance of Ledo-Burma road (presently Stilwell Road), with its reconstruction processes, i.e. recruitments of local people, the supply of construction items, men and materials etc. in the eve of Second World War.

The Post-war Notion

As the war was over all the responsibility of Stilwell Road were handed over by the Americans to the British authorities and it has been decided that military control on the road was no longer necessary and that the civil should take over control.[23] So, it is the duty to British to maintain the road as possible the needs of the people.

After the end of war and military need of the road, a consultation between A.F. Morley and the Chief of Staff with the department concerned was held about the strategic and political aspect of the road and reached the climax.[24] The maintenance of the links by road between Myitkyina and India and between Myitkyina and Tengchu has not necessity on the strategic or political ground.[25] For the strategic reason, the Myitkyina-Bhamo-Mongyu road should be maintained as an all-weather road.[26]

In the same time, the Burma Government office's conclusion was that the road need not be maintained as a link though the Burma Government were invited to consider that advisability of maintaining a small portion of it, for local use.[27]

But the importance of the road was not ended with the war. Because the road was too considered as a vital link among the China-Burma-India and the importance was earlier than war was broke out. In that consideration, the Indian Government decided that as long Government of Burma decided the maintenance of the road as a link with India could not justified and asked to External Affairs department to make an enquiry about the road to maintain and its approximate value.[28]

The Indian Government considered the Stilwell Road to be maintained in 12 feet width for the possible future emergence and handed over to Public Work Department.[29] The stretch of 27 miles onward from Lekhapani to Hell Gate Pass as a National Highway in a width of 12 feet and under the responsibility of the Political Officer of the area.[30]

While India and Myanmar getting independence in 1947 and 1948 respectively, both countries closed the historic Road, which was an economic and political lifeline for the region, for security purposes.[31]

From that time up to present, almost six decades the Road was closed and various parties raised their issues to reopen the Road. It was also demanded by international bodies also. A considerable amount of diplomatic efforts has been put into case out the past confronting issues and open-up new vistas of socio-economic co-operation and mutual trust between India and China. India and China have a rich historical, cultural and spiritual association, renewed in modern times after India and China reestablished their independent identities. As two most populous nations in the world, India and China shared a common aspiration to give their peoples a better and more prosperous future.

India's support of the pro-democracy movement in Myanmar caused strains in official ties between the two countries. In 1992, to mend some fences in their relationship, former Indian Foreign Secretary Sri J.N. Dixit and U Aye, a senior official in the Burmese Foreign Ministry took the initiatives in regularising the bilateral ties. U Aye's visit to New Delhi in August 1992, first by a Junta official since 1988 followed by Sri J.N. Dixit's visit to Burma in 1993, visit of U Nyuny, Dy. Foreign Minister of Burma to India with six days official programme in 1994, signing of MoU in 1994 to prevent illegal and insurgent activities, bilateral agreement to regularize and promote border trade in 1995, visit of Mr. K. Raghunath, Indian Foreign Secretary in 1999, visit of Myanmar's Cultural Minister in February 2002, SPDC Vice Chairman General Manug Aye in November 2001, Home Minister in November 2002, Dy. Foreign Minister in October 2000, Deputy Minister, Science and Technology, led a 15 member IT team to India in February 2001, visit of India's Chief of Army Staff General V.P. Malik in July 2000, Chief Naval Staff Sri Sushil Kumar visited in January 2001, Minister of External Affairs led a delegation to Myanmar to inaugurate the Tammu-Kallmyo-Kalewa Road and a remote sensing centre in Yangoon both established with the assistance from India, visit of External Affairs Ministry to attend India-Myanmar-Thailand Ministerial meeting on transport linkage held in April 2002 are some of the pointers to the active growing relations between the two countries.

Both countries come into some understanding and they signed some MoU for benefits of each other. Among this co-operation in Agriculture in April 1998, co-operation in Science and Technology in June 1999, credit lines of $ 15 million in November 2000, Cultural co-operation in January 2001 etc. are important.

The present development of their relation in political sphere brings some hopes on the development of communication too. Indo-Myanmar ties will help to develop the linkage with east *via* the nation which was in a halt for several years. The social-economic and cultural ties with the nations also brought some hopes of favour for both nations. Present Government also gives some hints to the opening of landlocks between these two nations which might extend the link up to other ASEAN nations. Already Manipur border is open for trade in Moreh and Tammu. The Stilwell Road's improvement work is in progress inside the Burma and a little portion remains to improve in bordering areas. In China, it is a Super-express highway which is in a six-lane way. It can able to connect the ASEAN nations, and North-east will become an economic hub, if it will become open, accordingly the Look East Policy/Act East Policy of Government of India.

Notes

1. R.N. Reid, 1966, *Years of Change in Bengal and Assam,* (London: Ernest Benn), p. 75.
2. Allchin, Bridget and Raymond, 1968, *The Birth of Indian Civilisation,* (London: Pelican), p. 23.
3. B.G. Verghese, 2004, *India's North-east Resurgent,* New Delhi, p. 10.
4. B.K. Barua, 1951, *A Cultural History of Assam,* Guwahati, p. 81.
5. H.K. Barpujari, 1963, *Assam in the days of Company,* 1826-1858, Guwahati, pp. 197-198.
6. John M'Cosh, 1837, *Topography of Assam,* Calcutta, pp. 7-13.
7. John M'Cosh, On the Various Lines of Overland Communication between India and China, Prof. Royal Geographical Society, London, Vol. V, (1860-61), pp. 47-54.
8. *Ibid.*
9. N.N. Acharyye, 1966, *The History of the Medieval Assam,* Guwahati, p. 61.
10. Edward Gaite, 1926, *History of Assam,* Calcutta, p. 51.
11. H.K. Barpujari, *op. cit.,* p. 197.
12. S. Hillaly, *op. cit.,* p. 37.
13. H.K. Barpujari, Francis Jenkins Report, *op. cit.,* p. xxiv.
14. H. K. Barpujari, *op. cit.,* p. 197.
15. A.J.M. Mills, Report on the Province of Assam, Appendix B, no. 275 of 1835, from Colonel Jenkins to A.J. Moffot Mills, para. 2.
16. F. Nos. K.W.2, Foreign, A. February 1983, Nos. 17-49, Report on the Hukoong Valley Expedition, Letter from J.F. Needham, pp. 189-49.
17. *Ibid.,* p. 74.
18. Record of the Engineer Section, Construction Service - Services of Supply, USAF, CBI and USF in IBT, March 1942—August 1944.
19. Lieut. Colonel Frank Owen, OBE, *Stilwell's Road,* Published at souvenir of Dihing Patkai Festival, 2000, p. 28.

20. S.N. Barua, The Stilwell Road, A content for the twenty-first century agenda, a paper presented at the second conference of NE Economic association held at Margherita College on February 2, 1999, p. 2.
21. S.N. Barua, *op. cit.*, p. 2.
22. Stilwell Diary, *History of Services of Supply, China, India, Burma Theatre*, Sec. III, Ch. IV, pp. 9-10.
23. F. No. 41-NEF/46, External Affairs, U.O. No. 3596-CA/46 dated 06.05.1946, NAI.
24. F. No. 41-NEF/46, External Affairs, Letter no. B/F&R, 2469/46, dated 21st August, 1946, NAI.
25. *Ibid.*
26. *Ibid.*
27. F. No. 41-NEF/46, External Affairs, U.O. No. 3596-CA/46 dated 06.05.1946, NAI.
28. *Ibid.*
29. F. No. 41-NEF/46, External Affairs, Letter No. Department of Labour, U.O. No. 651/VI/46 dated 16.02.1946, NAI.
30. F. No. 41-NEF/46, External Affairs, U.O. No. 3596-CA/46 dated 06.05.1946, NAI.
31. *Ibid.*

Bibliography

A.J.M. Mills, Report on the Province of Assam, Appendix B. no. 275 of 1835, from Colonel Jenkins to A.J. Moffot Mills.

Allchin, Bridget and Raymond, 1968, *The Birth of Indian Civilisation*, London: Pelican.

B.G. Verghese, *India's* 2004 *North-east Resurgent*, New Delhi, p. 10.

B.K. Barua, 1951, *A Cultural History of Assam*, Guwahati.

Edward Gaite, 1926, *History of Assam*, Calcutta.

F. No. 41-NEF/46, External Affairs, NAI, New Delhi.

F. No. 41-NEF/46, External Affairs, Letter No Department of Labour, U.O. No. 651/VI/46 dated 16-02-1946, NAI, New Delhi.

F. No. 41-NEF/46, External Affairs, Letter No. B/F&R, 2469/46, dated 21st August, 1946, NAI, New Delhi.

F. No. 41-NEF/46, External Affairs, U.O. No. 3596-CA/46 dated 06-05-1946, NAI, New Delhi.

F. Nos. K.W.2, Foreign, A. February 1983, Nos. 17-49, Report on the Hukoong Valley Expedition, Letter from J.F. Needham, pp. 189-49.

H.K. Barpujari, 1963, *Assam in the Days of Company*, 1826-1858, Guwahati.

John M'Cosh, *On the Various Lines of Overland Communication between India and China*, Royal Geographical Society, London, Vol. V, (1860-61), pp. 47-54.

John M'Cosh, 1837. *Topography of Assam*, Calcutta.

Lieut. Colonel Frank Owen, *OBE, Stilwell's Road*, Published at souvenir of Dihing Patkai Festival, 2000.

N.N. Acharyye, 1966. *The History of the Medieval Assam*, Guwahati, 1966.

R.N. Reid, *Years of Change in Bengal and Assam*, London: Ernest Benn.

Record of the Engineer Section, Construction Service-Services of Supply, USAF, CBI and USF in IBT, March 1942—August 1944.

S.N. Barua, The Stilwell Road, A content for the twenty-first century agenda, a paper presented at the second conference of NE Economic association held at Margherita College on February 2, 1999.

3

Road Construction in North-East Frontier[1] of India, 1820s–1830s

Santosh Hasnu

Warfare and road construction often go hand in hand. There is a close connection between war and road construction. In the North-Eastern frontier of India, the first Anglo-Burmese War (1824-26) saw frantic road constructions aimed at opening new line of communication for troop mobilisation. Military considerations created incentives for investment in 'public works' infrastructure. In the nineteenth century, the North-Eastern frontier witnessed the westward expansion by the Konbaung Burma that reached to the very borders of an equally expansionist British Bengal. In the process, the British East India Company (hereafter EIC) declared the strategic territories of Cachar and Jaintia as "protectorate state".[2] they were key to the security of British Bengal. The chapter analyses the major confrontation between the two growing powers – the British and the Burmese in the North-East frontier of British India. It focuses on how the "Company state" (referring to the policies and interventions of EIC) made roads during the war period. It argues that road construction provided a crucial site for the expansion of colonial state into North-east India. Roads built during the war period ensured the withdrawal of Burmese forces from Manipur, Cachar and Assam.

Even so, road building projects came to be folded into the "politics of access".[3] The state (EIC or the Company state) made efforts to regulate road building and conditions of access to the road. The *Company Raj* engaged and negotiated with local *Rajas* and hill chiefs in the process of showdown with the Burmese empire. When the war broke out in 1824, the Company officials obtained consent to build roads from the disposed *Rajas* of Cachar, Manipur and Jaintia through their territories. The *Rajas* of Cachar and Manipur also remained as caretakers of the newly constructed road connecting British Bengal with Upper Burma *via* Cachar and Manipur. On the contrary, studies on colonial 'public works' in colonial Orissa by Ravi Ahuja have shown how local rulers often opposed the construction of new roads including the obstruction of labour (Ahuja 2009: 169-171). Ahuja argues that the action was not only to prevent direct military intervention but also to avoid the annexation of state by the colonial overlords. In the case of colonial North-east India, British road building projects across the Khasi hills met with an armed response. In 1826, the EIC concluded treaties with the Khasi chiefs to have access to their hill territories and for road construction. The paperwork (treaties) was not, however, immediately translated into action, writes Gunnel Cederlof; both parties interpreted the agreements in conflicting ways that led to retaliation from both sides (Cederlof, 2013: 80). The Company officials also got involved in the internal politics of the hill chiefs which resulted in appropriation of territories. The conflict started when the Khasis in 1829 burnt down the sanatorium (bungalow) constructed by the Company Government at Nongkhlaw; released the convicts employed in the road construction and killed two British officers. The Khasi chief, Tirut Singh put-up a constant opposition to the colonial road construction project. In fact, construction and maintenance of roads, according to Chad Haines (Haines, 2004: 535-565), allowed for the expansion of state apparatuses and reoriented the social and cultural linkages of colonial subjects. The coming of roads in Khasi Hills was, therefore, perceived as facilitating all sorts of interventions by lowlanders (people of the Brahmaputra Valley); these signalled and aided colonial conquest of their hills.

Being promised to be reinstated on the throne, the disposed *rajas* of Manipur and Cachar were useful in organising labour force and coolies needed for the construction of roads. In addition, the chiefs were responsible for the upkeep and clearing of the roads. Thus, the *Company Raj* was able to subordinate the Nagas and Manipuris as 'coolies' along the road the Sylhet – Manipur. To build a road through the Khasi hills, Assamese convicts were put in force at the initial stage. With the coming of treaties, the Khasi chiefs were expected to lend a helping hand in the construction work. Therefore, this study examines how the historical events that occurred on the North-east frontier that aided the Company state to access and promote activities of road building which led to the colonial incorporation of the region.

British and Burmese Passages into 18th century Northern East of Bengal

To begin with, the first Anglo-Burmese War (1824-26) radically changed the balance of power in the North-Eastern frontier. The westward expansion of the Burmese kingdom led them into direct confrontation with the British EIC who had occupied eastern Bengal. It was a well-known fact that Burma long held keen interest in the internal affairs of frontier kingdoms in what is now called north-east region of India. Autonomous kingdoms in the North-Eastern frontier were in the eye of the storm when full-scale war broke out between Britain and Burma in the 1820s. These kingdoms were taken in under EIC 'protection' by means of treaties. The main ambition of the British EIC was to secure strategic positions for military purposes and to take control of the major routes of communication in the larger region, from the Brahmaputra to the Irrawaddy rivers.

Prior to the outbreak of Anglo-Burmese war, the region of Cachar and Manipur has been characterised by long distance trade through a network of monastic traders (Chatterjee, 2013: 46-56), movement of people and immigrants. The direct land route between Burma and British Bengal was only through Cachar and Manipur. The interconnecting waterways like the Surma, the Kushiara, and the Barak brought merchants and Brahmin priests from Bengal to Cachar and Manipur. The Manipuri chronicle records that trade mostly in elephants and horses existed between Manipur and neighbouring kingdoms of Tripura, Assam and Burma (Kabui, 1991:229). Further, constant trade and regular contact existed between Manipur and the Burma with several Burmese artisans being imported for the construction of temples (Kabui, 1991: 236). By the early decades of 18th century, Brahmin priests were already moving back and forth between the Burmese and Manipuri royal courts (Charney, 2011: 159-181). Eighteenth century Bengal has been characterised with political disturbance or crisis. The Mughal emperor Aurangzeb ordered that all *madad-i ma'ash* granted to Hindus be confiscated and the share of these grants steadily decreased in proportion to the Muslim share clear down to the reign of Alamgir III (Eaton, 1997: 263). Due to these socio-economic changes, the Brahmins move out in many directions. In a sense, we have movement of literati people looking for patronage in the princely states like Cachar and Manipur.

By the second half of eighteenth century, the Burmese had occupied the valley of Manipur. The *Raja* of Manipur, Jai Singh, sought the help of the British and a treaty of alliance was concluded between Manipur and the British EIC in 1762. The British Government dispatched Captain Verelst, the chief of the Company's administration in Chittagong, with a small British force march towards Manipur to aid the *Raja* in 1763.[4] Since Cachar was a gateway to Manipur, this political development led to the first contact between Cachar and the Company. Due to the difficulties of communication through the hills beyond the Barak, Captain Verelst failed to venture beyond Cachar, and he had to be recalled. However, by the latter part of the eighteenth century, private merchants and the East India Company had carried on trade with Cachar and the Company charged fees on trade across the border into Sylhet (Cederlof, 2009: 513-540). Europeans did a brisk business in elephant tusks which were bought in Cachar, markets thrived all around the Surma river and in the eighteenth century, the number of record market places increased overall in Bengal. Presumably, trade and markets also increased in Sylhet district which in 1790 had over 600 named market places (Ludden, 2003: 1-54).

Company officials often carefully recorded the existing roads and routes. James Rennell's map of 1765 clearly noted the routes from Khaspur (in Cachar) and Manipur.[5] Further, when Francis Buchanan Hamilton surveyed the Bengal region from 1807-14, he mentioned Khaspur (the capital of the Cachar kingdom) serving as a direct communication line between Bengal and Manipur.[6] When the *Raja* of Cachar threatened the Manipur *Raja* Jai Singh by killing his two messengers, the *raja* had a new road cut out to connect Manipur with Srihatta (Sylhet). The new road, as informed by one of the travelling priest to Buchanan, was passable for elephants, horses and bullocks.

When Lieutenant Thomas Fisher, another Company official who surveyed the Sylhet district and the newly conquered territories of EIC in 1821-22, one of his Manipuri informant who had widely travelled in Hindustan, informed him about the existence of three roads between Cachar and Manipur, and that the mountains over which the roads pass were covered with jungle of grass and bamboos cleared away in places by the Nagas. Fisher's informant remarked that:

> When Marjeet invaded Cachar, he crossed in 10 days, marching day and night and making every exertion to arrive before information of his march could reach Govind Chandra. There was formerly a road from Kaspoor to Munnipore which the distance was only 4 days journey but it is now overgrown with jungle.[7]

On the other hand, the Company's interest in the Khasi-Jaintia Hills was limited to trade in limestone. Many of the EIC officials were involved in private trade. For instance, Robert Lindsay, the first Resident Collector of Sylhet, made great fortunes in private trade mostly in limestone which came from the adjoining Khasi territories. Lindsay hinted on the proximity and the prospects of trade between two countries of Sylhet and China (Bhattacharjee, 4-5). In 1778, he recorded a caravan coming down from the interior of the mountains with several hill products including coarse silk from "the confines of China" (Bareh, 1987: 18). The Company officials expanded their activity in the mountains. During the late eighteenth century, however, conflicts over land arose; this involved Company officials, mountain Khasia *Rajas* and warriors, Bengal settlers and merchants, and private European merchants of various nationalities. The conflict that occurred on Pandua, concerning control over commercial routes and resources in highlands, however resulted into a war in 1790. David Ludden argues that the conflict resulted in drawing a new fiscal boundary which marked the northern limit of the British Bengal (2003: 1-54).

Therefore, in the pre-British period, there was instant of movement of people like merchants, Bengali Brahmins in the State of Cachar, Manipur and other neighbouring kingdoms. Further, a certain magnitude of trade networks existed across these regions and the EIC officials were engaged in trading activities. The circulatory flows can be explained as the result of commercial contacts and monastic networks in the past.

Pursuing a policy of annexation, the Kingdom of Ava (Burma) captured Manipur in 1817 and was on the verge of making inroads into Cachar by 1824. There was also a serious threat that the Burmese power would march towards Sylhet, which would afford greater facilities for extending its influence to the British territories. The British pre-empted Burmese advance by occupying Cachar. To check Burmese plan of invading Cachar and Sylhet, the British EIC earnestly opened negotiations with the rulers of Cachar, Manipur and Jaintia. The Cachar *Raja* Govind Chandra, fearing attack from the Burmese took refuge with the Company in Sylhet. Under these conditions, the *Raja* signed a treaty with the Company Government in 1824 assuming that he would be reinstated as Cachar *Raja* after the Burmese war. Overtures were also made to Manipuri princes and *Raja* of Jaintia. The takeover of Cachar under Company's protection was to enable the British to occupy the principal passes to the lowlands of British Sylhet and effectively opposed the Burmese to advance from Manipur.[8] According to Thant Myint-U, the British worried about losing this buffer zone that was important in their expansionist design. Therefore, they unilaterally declared Cachar and neighbouring Jaintia as protectorates and sent a force to halt the Burmese advance. It led to a confrontation between the two armies in Cachar. Coupled with a worsening situation along the disputed Arakan border, it eventually resulted in the British declaration of war on 5 March, 1824 against the Kingdom of Ava (Thant, 2001:18). Therefore, British recognition of Cachar and Jaintia as a "protectorate state" paved the way for the access and the incorporation of these principalities into the *Company Raj*. Thus, the First Anglo-Burmese War inadvertently brought the British colonial state in direct contact with several autonomous states such as Manipur, Cachar and Jaintia. This intervention further led the Company Government to invest in public works infrastructure like road construction.

Building Cachar-Manipur Road

By the end of 1823, the Burmese forces had advanced towards Cachar. The pretext given by the Burmese for inroads into Cachar was that they were coming to reinstate the rightful heir Govind

Chandra into the throne who was dispossessed by the Manipuri princes. To this claim, David Scott (the Agent to the Governor-General, North-East Frontier) replied that Cachar was under British protection and called upon the Burmese to withdraw their forces. There was no reply from the Burmese and soon afterwards a series of engagements took place on the frontier of Sylhet. A *vakeel* was despatched to Cachar by David Scott to procure more information regarding the position of the Burmese forces.[9] On the basis of intelligence provided by Naga informants, the *vakeel* reported that the Burmese forces had already occupied the eastern boundary of Manipur.

Being a "protectorate state", the Cachar *Raja* was asked by the Company Government to make every effort to clear the existing roads through his territory to Manipur.[10] Lieutenant Thomas Fisher (in-charge of Cachar Affairs) opined that the Cachar *Raja* Govind Chandra would not have any difficulty in putting the road in good order through Cachar from Badarpur to Kumranga.[11] Beyond the latter place, a large number of people were required to be employed in widening and improving the road to the Barak river, rendering the *nullahs* and hills passable for elephants and bullocks. The *raja* was also to prepare the other existing roads in case the Company army decided to march from several directions. The possibility of improving the road from Luckipur in Cachar across the Tippera hills connecting the Comillah district was also discussed. Fisher was of the opinion that with the exertions of Govind Chandra and the *Raja* of Tippera, the mentioned road might be rendered passable for men, elephant and horses. However, the Calcutta Government felt that opening-up this section of the road would facilitate the Burmese forces in entering the Company territories in Bengal.

As the war progressed, the construction of roads became more complex. The road between the Barak and Manipur passed through hilly terrains. This tract lie beyond the limits of Govind Chandra and his influence over the hill tribes was doubtful. Captain Dudgeon (in command of the Company troops) hoped that Nagas along the Cachar Manipur road could keep the road open if they work under the direct supervision of their respective chiefs.[12] There were five Naga chiefs between Cachar and Manipur, which the Company thought that their services might be readily engaged by the distribution of presents like woollen cloth, scarlet, liquors, tobacco, salt fish, onions and such articles much in request with them. Therefore, owing to the incapability of Cachar *Raja* to build roads beyond his territories, the Calcutta Council asked the Company officials to make an arrangement with Gambhir Singh or such person who had influence over the hill tribes to supervise the opening of the roads from the Barak to Manipur. It was found that there were men of influence with the Naga tribe who were willing to undertake the task. Tucker (Commissioner of Sylhet) recommended one Purbitha Singh for road construction, provided advance payments were made.[13] The choice was obvious, as Purbitha Singh was well acquainted with the localities between Cachar and Manipur, and possessed a considerable influence amongst the Nagas. In pursuance of removing every difficulty for the march of the Company troops, the Council authorized the employment of Purbitha Singh.

A Company official, Mr. Mathew, conducted a topographical survey to establish the best possible route from Sylhet. Surveying the line from Badarpur to Manipur, Mathew's report was to serve by giving best possible ideas of the practicability of making road to Manipur and the nature of the road as it stands and the difficulty to encounter the road.[14] He appeased the Naga chiefs with presents of a piece of red cloth in order to seek their help in preparing the road. Mathew suggested that, until some kind of road shall have been prepared, it will be impracticable for the transport of troops or the provisions necessary for its daily consumption.[15] The existing road or rather the route, according to Mathew, was nothing but a footpath where one could walk and that in many places with a great difficulty.

Decision for the plan of operations from British Sylhet frontier was taken in 1825. A considerable number of Company troops were to march from Cachar to Manipur and further towards Ava (Burma). Therefore, a road from Badarpur (in Sylhet frontier) to Banskandi (in Cachar plains) was speedily constructed by the pioneers for the passage of carriage cattle.[16] Further, H.H. Wilson writes,

> The country from Banskandi towards Manipur was a continual series of ascents and descents, the route being intersected at right angles by the ridges of mountains running nearly due north and south, the base of one springing from the foot of the other, with the intervention

> only of a mountain rivulet swollen into a deep and precipitous river after every shower: for the first thirty miles, also the sides of the mountains were completely covered with thick forest, the intervals between the trees of which were filled up with a network of intertwining reeds and brushwood, except where a narrow and often interrupted foot path wound through the labyrinth.[17]

When the First Anglo-Burmese War ended, the Company Government insisted on maintaining an open line of communication between Bengal and Burma. The British officials tried to persuade Gambhir Singh to contribute his best to the construction of the Cachar-Manipur road, for which the Cachar *Raja* too was expected to join.[18] On his restoration to the throne, the Cachar *Raja* received a loan of Rs. 2000 from the Government as financial assistance. Therefore, the Company Government used the opportunity to ask the *raja* to take responsibility in constructing the road in his territory in return of the favour made by the EIC. Hence, instead of paying tribute in 1827, Cachar agreed to construct a road from the plains *via* Cachar to Manipur facilitating easy communication with Sylhet.

However, the construction of the road required a large sum of money which the Cachar *Raja* could not afford due to depopulation caused by the Burmese war. To meet this demand for road construction, the *Raja* had imposed restrictions on the transit of articles, such as wax, ivory, silk and cotton between Manipur and Sylhet. The Cachar *raja* started to control the market, levying the toll taxes in order to finance the road construction. The *raja* began to behave like an entrepreneur. He averted his subjects from participating in the traffic of grain, which was exported in large quantities from Sylhet to Dacca and Narayaganj, and, therefore, compelled them to sell it to him so that the *raja* may reap the advantage of their labour.[19] The intention hurt the Sylhet market and the Company official complained of this act of the *raja* to the Calcutta Government. This act of the *raja* had also affected the British subjects who were engaged in the timber trade. This was seen as an unfriendly attitude towards the Company Government who had been providing protection to *raja* according to the terms of 1824 treaty.[20] However, David Scott perceived that there was no necessity for the Company Raj to place any kind of artificial check on the *Raja* of Cachar. By the end of 1820s, the British EIC was in the process of consolidating sovereignty in Cachar as the *raja* had no heir to succeed him. In 1830, the Cachar *Raja* Govind Chandra was murdered leaving no legitimate heir.

After the formal annexation of Cachar under Company dominion, Lieutenant Thomas Fisher was made the first Superintendent (in-charge of Cachar Affairs). The road project was now in the hands of Cachar authority. In a reply to the Company Government on the subject of the road through Cachar, Fisher reported that there were no roads in this territory, and bulky commodities were usually transported through the means of waterways. But, however, he writes that the paths for porters and laden elephants were open during the dry season over all the inhabited parts, which for temporary purposes might speedily be rendered practicable for bullocks.[21] Fisher requested for hundred convicts to be employed in the construction of road. The Company agreed to the request of Fisher but wanted the Judge and Magistrate of Sylhet to undertake the supervision of road building to a certain distance from Sylhet, when it might be taken up by Thomas Fisher under his supervision.[22] The coming of tea estates in Cachar region during 1850s increased the web of road networks. Roads were constructed connecting the tea gardens with the Barak river from where the tea trade was exported out of the region.

Along the Cachar and Manipur, Captain Grant (the Political Agent of Company Government at Manipur) pushed for drastic measures in building roads. Indeed the First Anglo-Burmese War had also made it clear for the need of constructing roads for the easy movement of troops across the hills lying between Cachar and Manipur. Captain Grant remarked that the road in the hilly tract lying remained unfeasible for cattle; however, the remainder of the way into Manipur was passable only for very lightly laden cattle.[23] Therefore, Captain Grant strongly suggested for construction of a passable road along this line which would not only facilitate the communication between Bengal and Manipur but also would be of immense advantage in the event of war. Adding up to military interests, roads in 1830s grew in importance for commercial routes as the trade had considerably increased. Manipur and Cachar formed the old overland routes for commercial networks between Yunnan province of China and Bengal. This existing trade network was interrupted by the war.

However, when the peace was restored by the year 1827, the trade was again resumed. The EIC sensed that Manipur could become the centre for trading activities between the northern province of Ava (Burma) and the north-eastern districts of British Bengal.

The Calcutta Government gave high precedence in taking necessary steps for constructing roads through Cachar to connect Sylhet with Manipur.[24] Against these measures, the EIC appointed Francis Jenkins and R. B. Pemberton to survey and suggest for the most suitable line for a road to be constructed between Cachar and Manipur. After travelling to these areas, they recommended that a permanent military road should be constructed between Cachar and Manipur. They suggested that a company of pioneer under a European Officer along with the assistance of a body of Nagas to be employed for the colonial road building project.[25] Being aided by the Company to regain his throne, the *raja* of Manipur was asked to provide the service of the Nagas and tools required for the road construction was to be delivered by Company Government. Both Jenkins and Pemberton also added that the portion of the road lying west of the Barak river should be kept in repair and cleared by the Cachar authorities and that on the opposite on eastern bank by the *Raja* of Manipur. This projected road was not only to make ease for the movement of British troops stationed at Sylhet in the event of any war but also to facilitate the trading activities. By the end of 1830s, the military department took over the task of constructing a road practicable for laden cattle between Cachar and Manipur. This was done more systematically with the service of sappers and miners who blasted the rocky pass in the hilly tracts.

The Making of Sylhet-Khasi Hills Road

When the Burmese forces occupied Assam in 1824, the EIC decided to despatch troops to displace the Burmese through Jaintia hills. In pursuance of access through Jaintia hills, David Scott negotiated with Ram Singh, the *Raja* of Jaintia. This led to the signing of a treaty between David Scott and Ram Singh by which the *raja* acknowledged the overlordship of the Company and promised to give his military support to the Company.[26] Through this treaty, the *raja* also agreed to the request for the construction of a road across the Jaintia hills. Thus, Company state began the construction of a road from Jaintiapur to Rahar in Nowgong, but the project was discontinued.[27] As around the same time, an idea developed to build a secure road for the direct link and communication between Sylhet and Assam *via* the Khasi Hills.

As contrast with Jaintia hills road, the projected road from Sylhet to Assam (*via* Khasi Hills) was considered vital as it would minimize the length of the march of the British troops. The EIC had been following the policy of non-intervention against the Khasi state since the conflict between the both in 1789-90. Since 1780s, there had been a constant cause of conflicts between Khasi chiefs, Company officials and private merchants resulting out of the competition to control the limestone quarrying and trade. David Ludden argues that these conflicts resulted in drawing of a boundary in 1791 defining the territorial control of the EIC and regulated mobility between Khasi mountains and Sylhet plains (Ludden, 2003: 1-54). Therefore, when the war broke out against the Burmese in 1824, David Scott began to persuade the Khasi chiefs and to bring them in constant communication with the Company Government by providing land as *zamindari* tenure in the lowlands of the Company territories.[28] By giving the possession of land at the lowlands and granting Khasis easy access to the markets below in Sylhet plains, Scott wanted to render the Khasi chiefs more dependent and subordinate to the Company Government. The idea behind these measures was to keep the Khasis in good humour and later seek permission for an access through their hills.

Soon, an application on the construction of road was submitted by the EIC to Chatter Singh, who during this period held the possession of lowlands of Barduar (a tract of land adjoining Nongkhlaw in Khasi hills). Chatter Singh readily consented to the request made by the Company State. In addition, Chatter Singh also agreed to grant all assistance required for the road construction.[29] However, his authority and power was contested in the hills by Tirut Singh. When Chatter Singh passed away, it led to the disputes from several claimants regarding the succession to the estates of the deceased. Rajjan Singh, a boy of only five years, brother and direct heir to deceased Chatter Singh claimed the tract of Barduar which was now under the Company's dominion. However, David Scott made it clear that he would not grant the tract of Barduar to any person whose authority was not fully recognised in the hills and who was not prepared to grant permission for constructing of a road.

In 1826, David Scott was asked to join at Nongkhlaw with the view of mediating between different contending parties. Through this assembly, Tirut Singh was elected as the *raja*.[30] Tirut Singh then applied to David Scott for granting him permission to rent the lands in the lowlands formerly held by Chatter Singh. Scott agreed to grant this only if the hill chief is ready to persuade the Khasis to consent for building a road through the hills to connect Sylhet and Assam.[31] When the Burmese troops retreated from Assam, the region came directly under the EIC's control. However, the Company forces were cut down by the Khasi hills separating Assam and Sylhet. The most urgent need was to construct a road across the hills. Accordingly, agreement was drawn with Tirut Singh who not only agreed to allow free passage for British troops but also promised to supply the required materials for the Company's road building project. Thus, the provision of road construction became an important feature of the treaties signed with different chiefs and *rajas*. Hence, Scott was instrumental in the rise of Tirut Singh, unites him into one single figure and helped in acquiring inheritance. In return of this favour made to Tirut Singh, Scott took the opportunity to get permission for the construction of a road and to provide a free passage for Company subjects through the Khasi state.

The Company Government decided to build a sanatorium (bungalow) and a police outpost in Nongkhlaw along with the construction of road. To put the public work into practice, David Scott requested for sending hundred Assamese convicts to be employed in making the road to Cherrapunji.[32] However, the agreement did not anticipate the fragmented political authorities existing in the Khasi hills. By the treaty, Tirut Singh looked up to the British as a potential ally and protecting his interests against any dispute with other hill chiefs. When a dispute broke out with Balaram Singh, the Raja of Rani (lying to the plains adjacent to Tirut Singh's territory), the Company Government refused to come to the aid of Tirut Singh. The hill Chief Tirut Singh came to understood the nature of the EIC and now wanted to keep away the colonial authority at a safe distance. This hill chief started to make a constant opposition against the improvement of the road and the labourers were discouraged from carrying on the work by the threats of massacre. At the same time, certain insolent speeches came to be circulated by the Bengalis employed in various capacities along the line of road that the hill Khasis would soon be subjected to taxes and unable to call their huts as their own.[33]

Roads, therefore, not only create access but facilitate movement of ideas. When the road was cleared and a bungalow was erected at Nongkhlaw, doubts arose in the mind of the hill Khasis that it was the beginning of more aggression on their hills (Baruah, 1985: 467). And, that the coming of the roads was seen as an infringement to the independent Khasi state. On April 4, 1829, a band of about 500 Khasis under the leadership of Tirut Singh attacked and massacred two English officers, Lieutenant Burlton and Bediengsfield, besides Indian sepoys at Nongkhlaw. They also burnt down the sanatorium and released the convicts employed in the road construction. The Khasi mobs at Nongkhlaw also rendered impassable by destroying the bridges and obstructed the road with trees.[34] Gunnel Cederlof argues that the attack had its roots in a treaty signed by Tirut Singh and David Scott in 1826, which included road construction through Nongkhlaw as part of a growing web of roads crossing the Garo, Khasi and Jaintia hills (Cederlof, 2013: 89). Therefore, road construction in the Khasi hills came to be entangled with the issues of independence of the Khasi state *vis-à-vis* the Company state and their resistance (which was 'rebellion' or 'robberries' to the Company state) to road projects.

Tirut Singh kidnapped the *amlahs* (Company local officials) and demanded a ransom (a certain sum of money) from the Company Government for their release. It was reported that the Assamese and perhaps Bengali settlers were in union with Tirut Singh. Captain White doubted that the slightest reverse would throw the entire hill territory with 'rebellion' as already the collection of revenues and coolies carrying provisions has been resisted.[35] Captain White added that "highway robberies" have started to commence in the territory. It was also reported that David Scott might also be detained at Cherrapunji for ransom. The EIC Government ordered the Company troops under Captain White stationed at Guwahati to march towards Nongkhlaw which was about 55 or 56 miles away. The road along this route was already constructed by David Scott which was quite practicable for horses but elephants could only get on with difficulty so all supplies to be carried-up by coolies.[36] David Scott was joined at Nongkhlaw by Captain White and accordingly started the temporary occupation of *raja* Tirut Singh's territory. Company declared a reward of a thousand rupees to apprehend Tirut Singh.[37] Thus, the politics of access dominated the ways in which EIC tried to suppress the armed response by the Khasis.

This politics of access (to the road projects) not only dwelled on the strategies of occupation of *raja's* territory, reward for apprehending *raja* and suppression of armed response, but also on other techniques. During the period of resistance and when Tirut Singh was rallying around him, EIC opened negotiations and agreements with the other Khasi chiefs. The Company Government tried to bring the different Khasi chiefs under their influence by granting plot of lands at lowlands lying in Sylhet frontier. For instance, the EIC ceded a portion of land at Pandua in Company territory to *raja* Dewan Singh of Cherrapunji for the purpose of enabling him to establish a *bazaar*.[38] In return for this favour made by EIC, Dewan Singh granted permission for the construction of road and also bestowed a site at Cherrapunji for the construction of sanatorium. The acts of violent behaviour, however, still continued by the inhabitants of different villages who interrupted the line of communication with Sylhet.[39] Between 1829 to 1833, there were numerous resentment against the EIC. Until in 1833, when Tirut Singh was brought into terms with the Company Government and sent in exile as political prisoner to Dacca (Bareh, 1987: 301-302). To render the possibility to occupy any other post in the interior of the Khasi mountains, the construction of road was seen important which was to be accomplished by two companies of pioneers and assisted by the convicts.[40] Accordingly, the British built roads to control the Khasis and maintain access through their hills. Fines that were imposed and collected from the rebelled chiefs and villages were utilised for the construction of public works such as roads. In some cases, the Khasis agreed in the way of payment in labour. And in other cases, the Khasi chiefs agreed to make a passable road in their respective territories.[41] For instance, out of a sum of five thousand imposed on Chand Manick by David Scott, the chief paid only four thousand. And in return for not able to pay the full amount, this chief agreed to make a passable road from Cherrapunji to Myrung with the aid of convicts.

In the 1830s, when the Company's position was much stronger in the hills, the EIC started to plan out for a systematic opening-up of the Khasi hills. For this purpose, Jenkins and Pemberton were sent to survey and were required to give their opinions for the constructions of the road in the Khasi hills. After the survey, both Jenkins and Pemberton asserted for the viability of making a practicable road in the hills. To complete this colonial road building project, both of them recommended for the employment of Assamese convicts and a party of Khasis to assist in the road construction.[42] They also recommended the appointment of an Engineer officer to superintend the construction of roadwork and to superintend the erection of bridges. Fines and rents collected from Khasi villages and from the Chiefs were decided to be utilised for the local improvement of the roads and bettering of the main branches of roads that connect Assam and Sylhet.[43] In the 1830s, the road construction project from Sylhet and Assam was further taken over by the military department. With the takeover of the road building by the military department, the period of opening a good road with iron bridges and wooden bridges was effectively executed in the Khasi hills to connect Sylhet with Guwahati (Assam).

To retrospect, the First Anglo-Burmese War (1824-26) became a significant event for colonial intervention in the North-East Frontier and for the road construction. Strategic imperatives dominated the EIC's decisions to build overland routes connecting Sylhet with Burma *via* Cachar and Manipur and building secure link between Sylhet with Assam *via* Khasi hills. In that sense, roads became a part of war legacies. Road building activities were a strategic component that contributed to the success of the British against the Burmese, as well as the colonial conquest of the North-east region. To consolidate its power in the region through further conquest, EIC required evolved strategies to regulate and access road politics of access. In brief, our study shows that historical analysis of roads is an important field of historical enquiry to understand state and empire formations in a region.

However, colonial road building provoked varying responses in different parts of the North-east India. Whereas it managed to build roads along Cachar and Manipur, it made with armed resistance in the Khasi hills. Road construction through the heart of Khasi hills met with denial of access by the Khasis to their territories. Cachar and Manipur *Rajas* viewed the British as an ally against Burmese threat but Tirut Singh saw the British as a threat rather than Burmese. Local rulers, caught-up in different political circumstances, had different perception of threat. This chapter, therefore, was an attempt to highlight how the war-time roads of the colonial state were laid out. Construction and improvement of roads continued even after the Burmese war ended, levelling out certain sections and erecting more permanent bridges. The commercial aspects of the roads meant to extend trade and open-up new markets in the region began during the 1830s.

Notes

1. The term 'North-East frontier' was an administrative unit under an Agent to the Governor General. The British EIC defined 'frontiers' from the perspective of Calcutta.
2. A 'Protectorate State' was a native princely domain which enjoys the military protection of the EIC in exchange for bearing the cost of Company forces.
3. The phrase 'politics of access' has been used by Mahnaz Ispahani to express how the State regulates condition of access to routes. *See*, Ispahani, *Roads and Rivals*, p. 7. More recently, Lipokmar Dzuvichu used this idea in studying the road building project in colonial Naga Hills, *See* Dzuvichu, pp. 473-494.
4. Pemberton, 1835, *The Eastern Frontier of India*, Baptist Mission Press, Calcutta, 1835, pp. 19-20.
5. Rennell description of routes is mentioned in Lieutenant Thomas Fisher's Memoir of 1824. National Archives of India (hereafter NAI), New Delhi, Foreign & Political Department, Consultation 18th June No. 59, 1830, Extract from Lieutenant Fisher Memoir in Cachar dated 24th November, 1824.
6. Hamilton, *An Account of Assam*, pp. 83-87.
7. NAI, New Delhi, Foreign & Political Department, Consultation 18th June No. 59, 1830, Extract from Lieutenant Fisher Memoir in Cachar dated 24th November, 1824.
8. Wilson, *Narrative of the Burmese War*, pp. 19 & 22.
9. NAI, New Delhi, Foreign & Secret Department, Consultation, 20th April, 1824, No. 5.
10. NAI, New Delhi, Foreign & Secret Department, Consultation, 30th April, 1824, Nos. 33-34.
11. NAI, New Delhi, Foreign & Secret Department, Consultation, 14th May, 1824, Nos. 8-10.

12. *Ibid.*
13. NAI, New Delhi, Foreign & Secret Department, Consultation, 11th June, 1824, Nos. 21-23.
14. NAI, New Delhi, Foreign & Secret Department, Consultation, 31st December, 1824, No. 7.
15. NAI, New Delhi, Foreign & Secret Department, Consultation 7th January, 1825, Nos. 22-23.
16. Wilson, *Narrative of the Burmese War*, p. 133.
17. Wilson, *Narrative of the Burmese war*, pp. 133-134.
18. NAI, New Delhi, Foreign & Secret Department, Consultation, 23rd November, 1827, Nos. 37-38.
19. NAI, New Delhi, Foreign & Political Department, Document No. 81: To the Chief Secretary to Government, Fort William from Tucker, Commissioner, Sylhet Commissioner Office, 18th February, 1829.
20. NAI, New Delhi, Foreign Political Proceedings of 7-21st May, 1832, Document No. 84, Chief Secretary Lwinton, Fort William 30th May, 1829 to the Agent to the Governor-General, North-East Frontier.
21. NAI, New Delhi, Foreign & Political Department, Consultation 12th November, 1832, No. 43, To Robertson, Agent to Governor-General, North-East. Frontier from Lt. Fisher, Cachar, 10th October, 1832, Reply to a letter on the projected road from Sylhet to Banskandi.
22. NAI, New Delhi, Foreign & Political Department, Consultation 12th November, 1832, No. 44.
23. NAI, New Delhi, Foreign & Secret Department, Consultation 23rd January, 1832, Nos. 6-9.
24. NAI, New Delhi, Foreign & Political Department, Consultation 17th September, 1832, No. 17.
25. NAI, New Delhi, Foreign & Secret Department, 2nd July, Nos. 12-13, 1832: To Robertson, Agent Governor-General, North-East Frontier from Jenkins & Pemberton, On Special Survey Duty, Calcutta 19th June, 1832.
26. Aitchison, C.U., 1931, *A Collection of Treaties, Engagements and Sanads Relating to Jammu & Kashmir, Sikkim, Assam and Burma*, Vol. XII, pp. 118-119.
27. 'Chirra Punji', *Asiatic Journal*, Vol. XXI, 1836, p.17.
28. NAI, New Delhi, Foreign & Secret Department, Consultation 2 April, 1824, Nos. 16-17.
29. NAI, New Delhi, Foreign & Secret Department, Consultation 26th June,1829, Nos. 1-4.
30. *Ibid.*
31. Mackenzie, Alexander, *The North-East Frontier of India*, p. 221.
32. NAI, New Delhi, Foreign & Secret Department, Consultation 17th October, 1828, Nos. 4-5.
33. Pemberton, *The Eastern Frontier*, p. 232; See also Mackenzie, *The North-East Frontier of India*, p. 222.

34. NAI, New Delhi, Foreign & Secret Department, Consultation, 1st May, 1829, No. 42.
35. NAI, New Delhi, Foreign & Secret Department, Consultation, 8th May, 1829, Nos. 9-10.
36. NAI, New Delhi, Foreign & Secret Department, Consultation, 8th May, 1829, Nos. 11-12.
37. NAI, New Delhi, Foreign & Secret Department, Consultation, 5th June, 1829, No. 6.
38. NAI, New Delhi, Foreign & Secret Department, Consultation, 4th September, 1829, No. 18.
39. NAI, New Delhi, Foreign & Secret Department, Consultation, 23rd October, 1829, Nos. 14-15.
40. NAI, New Delhi, Foreign & Political Department, Consultation, 13th August, 1832, Nos. 66-68.
41. NAI, New Delhi, Foreign & Political Department, Consultation, 22nd May, 1834, Nos. 78-79.
42. NAI, New Delhi, Foreign & Political Department, Consultation, 15th October, 1832, Nos. 112-114.
43. NAI, New Delhi, Foreign & Political Department, Consultation, 22nd May, 1834, No. 78.

References

Ahuja, Ravi, 2004. "The Bridge-Builders: Some Notes on Railways, Pilgrimage and the British Civilizing Mission in Colonial India". In *Colonialism as Civilizing Mission: Cultural Ideology in British India* (eds.) Harald Fisher-Tiné, Michael Mann, pp. 95-116. London: Wimbledon Publishing Company.

Ahuja, Ravi, 2009. *Pathways of Empire, Circulation, 'Public Works' and Social Space in Colonial Orissa, c. 1780-1914*. Hyderabad: Orient Blackswan.

Bareh, Hamlet, 1987. "Khasi-Jaintia State Formation". In *Tribal Politics and State System in pre-colonial Eastern and North-Eastern India,* (eds.) Surajit Sinha, pp. 261-305. New Delhi: K.P. Bagchi & Company.

Baruah, S.L., 1985. *A Comprehensive History of Assam*. New Delhi: Munshiram Manoharlal Publishers.

Bhattacharjee, J.B., 1997. *Cachar under British Rule in North-East India*. New Delhi: Radiant Publisher.

Bhattacharjee, J.B., 2000. *Trade and Colony: The British Colonisation of North-East India.* Shillong: North-East India History Association.

Bhuyan, S.K., 1949. *Anglo-Burmese Relations 1771-1826*. Guwahati: Lawyer's Book Stall.

Cederlof, Gunnel, 2009. "Fixed Boundaries, Fluid landscapes: British expansion into Northern East Bengal in the 1820s". *Indian Economic and Social History Review,* 46: 4, pp. 513-540.

Cederlof, Gunnel, 2013, *Founding an Empire on India's North-Eastern Frontiers 1790–1840*. New Delhi: Oxford University Press.

Charney, Michael, 2011, "Literary Culture on the Burma-Manipur Frontier in the Eighteenth and Nineteenth centuries", *The Medieval History Journal*, 14, 2, pp. 159-181.

Chatterjee, Indrani, 2013. *Forgotten Friends: Monks, Marriages, and Memories of North-east India*, New Delhi: Oxford University Press.

Deloche, Jean, 1994. *Transport and Communication in India prior to Steam Locomotion*. New Delhi: Oxford University Press.

Dzuvichu, Lipokmar, 2013. "Roads and the Raj: The Politics of Road Building in Colonial Naga Hills, 1860s–1910s", *Indian Economic and Social History Review*, 50, 4: 473-494.

Dzuvichu, Lipokmar, 2014. "Empire on their Backs: Coolies in the Eastern Borderlands of the British Raj", *International Review of Social History*, No. 59, Special Issue, pp. 89-112.

Eaton, Richard M., 1997. *The Rise of Islam and the Bengal Frontier 1204-1760*. New Delhi: Oxford University Press.

Haines, Chad, 2004. "Colonial Routes: Reorienting the Northern Frontier of British India", *Ethnohistory*, Volume 51, Number 3, pp. 535-565.

Hamilton, Francis, 1963. *An Account of Assam 1807–1814*. Guwahati: Department of Antiquarian Studies.

Ispahani, Mahnaz Z., 1989. *Roads and Rivals: The Politics of Access in the Borderlands of Asia*. London: I.B. Tauris & Co. Ltd. Publishers.

Kabui, Gangmumei, 1991. *History of Manipur, Volume One–Pre-Colonial Period*. New Delhi: National Publishing House.

Ludden, David, 2003. "The First Boundary of Bangladesh on Sylhet's Northern Frontiers", *Journal of the Asiatic Society of Bangladesh*, 26 June, pp. 1-49.

MacGeorge, G. W., 1894, *Ways and Works in India*. London: Archibald Constable & Company.

Medhi, Shyam Bhadra, 1978. *Transport System and Economic Development of Assam*, Publication Board, Guwahati, Assam.

Myint-U, Thant, 2001. *The Making of Modern Burma*. United Kingdom: Cambridge University Press.

Pemberton, R. B., 1835. *The Eastern Frontier of India*. Calcutta: Baptist Mission Press.

Verghese, K.E., 1976. *The Development and Significance of Transport in India 1834-1882*. New Delhi: N.V. Publications.

4

Politics of Road Construction in the North-East India

Prospects and Challenges

Mayengbam Nandakishwor Singh

Human Civilization and Necessity of Road Construction

In the human history, many civilizations have perished, some have survived and many are still flourishing. Human civilization never cease to travel in its own course and it is an ongoing journey. In all these unending continuation, connections among human civilization have been bridged by the road connectivity. Disregarding of the imminent fate of some civilization, the connection among humans largely ensure that continuity of human civilization as a whole progress. Human beings can never live apart, isolated from fellow human beings in stage of human civilization. It is natural for the human species to be in the connection with others for varied tangible and intangible needs and it is a fact that human beings alone cannot fulfil all what they need for the complete life. The fast imposing trend of globalization is the classic proof of this human trajectory. Evolution of human civilization is shaped by the connectivity among human beings and it is the roads of connectivity that largely enable it.

Road, a way on land, is meant for providing connections among humans. Road construction is so essential for varied reasons and its necessities have been burgeoning in each passing time. It is true that as long as human thrive to excel in all spheres, it is inevitable to seek the roads towards their destination. Since exploring ways to fulfil the desire for human excellence is the call of the inner human nature, road construction has ever been one of the utmost drives. It is the tendency in human beings to stay connected and thus, development of roads becomes so indispensable since the dawn of human civilization. Human beings cannot fulfil their natural needs being in isolation for long and it is the natural tendency for humans to complete the inner desire by being connected with others.[1] In the process, excavating ways to connect with other becomes so compelling that human always vigorously build roads big or small throughout the ages. There is big political reason behind the construction of roads in any given geographical territory. Looking at the ancient habit of humans, it is known that they had the tendency of domination over others and in many literature, it is characterized as the hegemonic character to reign over others.[2] Kings sought to control over other kings and in their missions, they build roads to pass on to the unknown territories. Kings and emperors mostly engage in bloody wards and battles with others kings for varieties of ends. Some of the notable reason are the desire to annex the territory so as to expand the sphere of control and to gain material benefits. For doing so, roads were built as the first necessary step towards that target territory.

In the modern times, various countries also practise imperialism in which they directly or indirectly control over other countries. In this act of imperialism, roads provide the connectivity and colonial masters explore the ways to control other countries. Political and economic gains drive the imperialist countries to follow this path fiercely. And for furthering their multiple interests, imperialist countries always put in action of their imbibed culture of road constructions wherever they venture. In fact, it is the imperialist countries that explore countless virgin territories and present their existence to the world at large. Taking the context of third world counties, it is the imperialist masters who steered in making roads in countless unfriendly terrains and even in the contemporary times, we find ample remnants of it. Even if there is indelible element of large scale exploitation, construction of roads by the imperialist countries still serve as the boons for the undeveloped and developing countries in the modern times around the world.

In this age of rapid globalization, almost all the countries fiercely pursue economic development. Economic development can be brought in two broad spheres: one is domestic sphere and other is in external sphere. For bringing economic development, construction of infrastructures, planning, tapping the natural resources and making use of technology are the ways. In other words, for bringing an all-round development in domestic circle, providing connectivity is the first and immediate step and roads provide these prerequisites. Without the proper road connections, the above mentioned requirements can never be actualized. And in the international sphere too, building road connections with the other countries, particularly with the neighbouring countries is the way forward. Any country cannot conduct its economic activities without proper road connections and thus it becomes the immediate step to embark on road constructions. Once, the road construction is activated, other facets of development too naturally come along.[3]

In today's geo-politics, no country can stay aloof for its own security. Countries around the world seek for strategic alliances with other countries. Several diplomatic ties are build-up for their security and other purposes like the collaboration in the field of arms, making new technology, sharing intelligence and so on. For security purpose, border roads are constructed and countries invest huge amounts on them. Making road inroads into the strategically located areas within or outside the country becomes one of the modern tools of successful geo-politics today.

Challenges of Road Construction in the North-East India

Geographical terrain is one of the major challenges facing the construction of roads in the region. It is unequivocal fact that road construction in the north-east region is one of the daunting task. It is challenging owing to several factors. Geographically, the region is located in the vast uneven, hilly and difficult terrains. Construction of road in this region always calls for special expertise and manpower. This natural geographical terrain of the region always plays as the biggest ominous factor in any road construction thus requiring longer years of labour along with huge financial back-up. Obviously, the cost of road construction in the region is higher and more expensive. As the region inherently encounters many disadvantages in all the political bargaining since the inception of present Indian political system, road construction has been constantly sluggish and not moving fast.

Political quagmire in the region poses the most everlasting challenges in the road construction in the region. The politics in the region is never steadily stable ever since the British left the region. This political instability always derails any kind of road development in the region. Most of the ruling parties in the region have to engage with the challenges posed by the powerful and influential militants active in the region. Amidst all the odds, many governments of the states cannot take-up any road construction as part of the development independently. Construction of roads in the city or state capital areas can be visible; but the fate of road construction in the interior parts of the states is always folded. Through a network of complex nexus, road construction in the interior parts is frequently checkmated.[4]

The model of economy dependency is seriously kneading the road construction in the north-east India. The region does not have an independent economy of its own for years. It can be because of the unequal political arrangement India has adopted since its independence. The region has a kind of dependent economy that it has to bank upon the centre for any financial assistance for any development. And since the region is more or less poor economically, this severely affects any plan for road construction in the region. On the whole, region's indigenous industries are not properly promoted. On the top of it, new policy for road development hardly reaches the region in time.

Corruption is another factor that really plays as the bottleneck of the road construction in the region. Corruption is exercised in multiple levels both in the Central and State level. Even if either the Central or State Governments announce the construction of any road in the region, it takes long time to actually materialize them. Today, as corruption is the unofficial yet acceptable norm in all strata of the society, no significant action advances independently of corruption. Thus, any plan for road construction is not immune to this disease of corruption and nothing cannot be proceeded in a fair manner. Even if budgetary allocation for the road construction in the region shows definite number in pen and paper, actual amount utilized for the actual road construction always reflect different picture.

Another simmering challenge with regard to the road construction is based on the fear psychosis that people of some specific areas hold. The fear is that proper road connection will drastically change the demographic equation in that area. Small communities may get swamped in the sea of larger communities and they will be disadvantaged in all fronts. As a preventive measure, complex legal or customary restrictions are imposed in several areas of the region.[5] In such areas, construction of roads is a big challenge for the government as permission for the land and the forest are to be legally sought for.

Politics of Road Construction in the North-East India

Historically, culturally, racially, geographically, north-east India is quite different from the rest of the India. The region has its own share of history disconnected with the rest of the country and it has the most diverse rich cultures. Moreover, the region has distant connections with the rest of the country. But one single connecting factor is the British rule which for the first time put together the region with the other parts of the country in the political map. It is after the independence of India that the region became formally part of the present existing political system. Since 1947, the political alliance between the region and the centre has been one of the love lost relations in south Asia. The road construction in the north-east region has its historical roots in this context.

The region is strategically located in the South Asia and South East Asia and there is serious security question in this region. Looking from the Indian point of view, the region is bordering with many significant countries. Numerous challenges lie ahead of any unplanned road constructions in the region. A knee jerk perspective regarding the road construction would be that by making roads, it can ensure security of the region. But this is a narrow judgemental view. As the region's being a part of India is ever an ongoing contestation, it is not feasible from the strategic point of view to build roads in the region. One of the unsaid yet practical fear is that modern road connectivity in the region will result in pushing it closer towards its immediate neighbours. This will pose a serious challenge to the present political existence of the country. For instance, China's active or clandestine support to the insurgencies of the north-east region which are battling for the secession from India will be further emboldened.[6] This dynamic may escalate the already vexed situations. In addition, the region has more close affiliation with its immediate neighbours in terms of racial features, cultural affinities, and economic advantages etc. So, the practical politics in this context is that construction of roads should not amount to tamper with the security of the country.

Owing to the present contestable political alliances, the region is never treated to be a part of so called 'mainstream India' psychologically. Whenever there is any planning for the whole country, the region gets slack attention and it receives a meagre budgetary allocation for the construction of roads. Financial allocation for the region is all time negligible. Often, equal importance is not given in the region and thus even after many years of independence, there are areas in the region which is yet to experience the modern lifestyle. Focus has always been on the developing the Indian main cities to be in the world map. One of the true, yet unacknowledged views is that the region is kept for security as a "buffer zone." The purpose is to protect the mainstream lands from any possible assails from the powerful neighbouring countries. So, active promotion of the region in all aspects is never a priority for the centre as it has some repercussions.

The region is an insurgency hit area and it is not in the interest of any ruling political dispensation to advance road construction. The Central Government's point of view is that construction of roads in the region will pave the way for the region to be closer with south Asian and south East Asian countries. This in turn will open more avenues for the insurgencies operating in the region to garner more support from its hostile neighbours. In the long run, it will severely explode the legitimacy of India's political claim over the region. The best pragmatic way is ensuring the region to move forward in a slow pace while pushing the process of mainstreaming. Parallel to this line of thinking, there is a politics in the road construction related to any ruling political dispensation of the states in the region. Since insurgency is always a challenge to the smooth functioning of any ruling party in the states, they too toe line adopted by the Central Government. And since the Central Government largely control the financial assistance in the region, State Governments do not have the sound and independent economic power to build up roads. Even if any promising programme of road construction in the region is taken-up from time-to-time, unusual delay in the completion is the visible norm.

Interestingly, there is also another interesting dimension about the road construction in the region. All the insurgencies functioning in the region have different aspirations. Some are fighting to secede from the Indian union, while many are fighting within the Indian Constitution for separate statehood. Among them, some of them try to showcase the legitimacy of their demands based on the underdevelopment or backwardness of the areas of their influence. In this scenario, they often hold either a dominant ruling community or the ruling government at the state responsible for having marginalised their cause for long. In order to validate their claims, they intentionally block any development programmes in the areas. Funds meant for development of roads and other purposes are regularly siphoned off by making illicit nexus with several agents.[7] In one way, they embezzle and extort all development funds and thus making it one of their main financial source. Road construction in their areas of influence poses a serious threat to their activities, and thus it becomes pertinent for them to choke the plans at any costs. It is because if roads are constructed in the areas, it will be double whammy for these insurgency groups. This will invalidate their demands of underdevelopment and this also will direly weaken their activities as Central or State forces can have easy route to their strongholds.

Prospects in Priming the Road Construction in the North East India

Proper road construction in the region will do wonders in many fronts. Developing proper road connectivity within the state or among the states in the region will only result in the economy prosperity. It is true that lack of road connectivity slows down the economic progress of the region. Once economy is developed in the region it can slowly snowball its economic activities beyond the region in the long run. All the small scale enterprises can also boom in the process. Even construction of the roads in the region will generate many jobs for the local people. On the whole, this may gradually bridge the widening gap between the urban and the rural.

With the coming the "look east policy," road construction in the region is an immediate matter of attention. The proposed policy of the look east policy is to have economic ties with the South East Asian countries which are developed and share close affinities with the region. In order to reap the benefits of the coming look east policy, proper road construction calls for proper planning. The region's possible benefits from this policy will largely depend on the road connectivity it may develop at present or in the near future. The region may not benefit itself fully from this policy sans road connectivity.

One of the most salient faces of the north-east region is the prolonged political uncertainty. Political instability can be seen in two broad perspectives. One is the long existing ethnic conflicts in the region which affect the political conditions in the region. Another is the tussle between the Centre and groups demanding for the sovereignty in the region. A close observation reveals that these two elements are intertwined. Then what is the connection between road construction and the political instability in the region? Basic causes of the ethnic conflicts have their roots in the long misunderstanding deriving out of the centuries old absence of intermingling with one another. In the region, different ethnic groups have been inhabiting in isolation. In the past, there was no proper means through which they could hobnob freely with other groups. This is still continuing in many parts of the region and this can be reasonably mitigated once roads are constructed thereby ensuring the interconnectedness among the ethnic communities. This process will help the ethnic groups to understand each other and thus abate their common misunderstanding eventually.

Even in the national-level, having proper road construction will have significant impacts. Due to the lack of sufficient road connectivity in the region, huge chunk of its people are constantly deprived of basic health facility, good education, access to modern technology, equipment and so on. When roads are efficiently constructed in the region, these acute problems can be significantly reduced and the outcome will be that basic education and health care will reach to the larger section of people. This will enlighten the citizens about their rights, privileges, and opportunities and so on. And eventually, once the people of the region becomes sophisticated and vigilant in all fields, they can have better stands to bargain politically with the centre for whatever is due to people of the region.

Concluding Remarks

Road construction intrinsically goes with the progress of human civilization. It has social and cultural impacts apart from the political and economic angles. In the context of the north-eastern region of India, road construction is slowed down because of the various economic, political and geographical reasons. Many of the factors hampering the prospects of road construction can be observed both in its internal as well as external realms. Both state and non-state actors are equally responsible for the derailment of the road construction in the region. In spite of many possible benefits which can be generated, the region is not in the position to harvest the fruits. The north-east India immediately needs to kick-start the programmes for road construction in order for it to be on par with the rest of the world. In fact, lack of infrastructure and development is the main cause of the acute problems in the region. Probing closer, this can be achieved through the construction of roads and there is a natural interdependency between the road connectivity and building infrastructures.

Notes

1. Such notion about human nature had been long theorized by Aristotle when he claimed humans to be social and political animal beyond which men cannot fulfil a good life (Aristotle, 1992: 59-60).
2. Thomas Hobbes's idea of pessimistic human nature in the pre-political state of nature in his *Leviathan* can be one refreshing point of reference in this regard.
4. Urban areas are in the nose of the State Government and state machinery has easy access; while far off remote areas are under the influence of insurgent groups in the north-east region and this is one significant cause.
5. In some states of north-east, such as Nagaland and Mizoram, there are some constitutionally protected customary tribal laws, which either Central or State Government cannot overrule at will.
6. China and India are two rising rivals in Asia and anti-India elements often find their support from China. In the region, it is obvious that China lend the logical and strategic support to various insurgent groups.

7. Reliable sources and intelligent reports clearly point out that insurgent group like NSCN (IM) habitually carries out such dastardly activities, particularly in some hilly districts of Manipur, to further its agendas.

References

Aristotle (2000): *The Politics*. London: Penguin.

Das, Sailen D. (2015): *North-east India in India-China Relations*. Guwahati: DVS Publishers.

Hazarika, Sanjoy and V.R. Raghavan (2011): *Conflicts in North-east: Internal and External Effects*, New Delhi: VIJ Books.

Hobbes, Thomas (2002): *Leviathan*, London: Penguin.

Kaul, Manmohini (2015): *India's Look East to Act East Policy*. New Delhi: Pentagon Press.

Lapp, John A. (1916): "Highway Administration and State Aid", *American Political Science Review*, 10 (4): 735-38.

Patterson, August (2016): *The Buffer Zone*, North Charleston: Create Space Independent Publishing Platform.

Raatan, T. (ed.) (2008): *Encyclopaedia of North-East India*. Delhi: Kalpaz Publications.

Roy, Shibani and S.H.M. Rizvi (2011): *Tribal Customary Laws of North-East India*, New Delhi: B.R. Publishing Corporation.

Ruban, Michel (2002): *Quality Control in Road Construction*, Tokyo, A.A.: Balkema Publishers.

Themis, Mary Olympia (2016): *The Code of Human Civilisation*, Bloomington: Xlibris NZ.

5

Road to Ruin

An Inquiry into the Prevalence of Corruption in Road Building in India

Dibyajyoti Dutta

"When plunder becomes a way of life for a group of men in a society, over the course of time they create for themselves a legal system that authorizes it and a moral code that glorifies it" —**Frederic Bastiat**

Introduction

The Road infrastructure is central to development. Roads facilitate societal connections and cultural linkages. Since the colonial period, road making has assumed centre stage in the developmental discourse in India. Road networks were established for administrative conveniences during the colonial regime. But with the indépendence from the colonial clutch, policies of road making were directed towards economic development and better social connections. Indian road network today is the second largest in the planet. Well planned, properly maintained, and safe roads are critical for economic growth and overcoming poverty in developing countries. The roads sector has also been a major target for development financing. Between 2000 and 2010, the World Bank (the most prominent finance provider of development projects in developing countries) committed close to 56 billion US dollar for road construction and maintenance.[1] The road making has been a constant and priority based programme of most of the developing countries including India. Roads are seen as the driving force of economic growth and as such it remains at the core of development strategies of the Government since the introduction of the First Five Year Plan. Apart from ensuring greater connectivity, the most significant and visible change that road infrastructures have brought is in the agriculture and allied sectors which contributes 16 per cent to the GDP during the financial year of 2014-15. Better road connectivity has the potential of accelerating agricultural production, guarantee sustainable market system and generation of increased agricultural income. Though road infrastructure holds tremendous economic prospects, but everything is not right with the implementation of road building programmes in the country. A plethora of scams have taken place in the country as has been periodically brought to light by different agencies including the highest auditing institution of the country, the Comptroller and Auditor General of India (CAG).[2] It is nonetheless true that the corruption in road making is not something unique to India. Road projects around the globe remain plagued by fraud, corruption, and collusion. A Transparency International poll ranked construction as the industry most prone to corruption and a survey of international firms revealed that companies in the construction industry were more likely than firms in any other sector to have lost a contract because of bribery. World Bank-financed projects are also not immune from this phenomenon. Roughly one-fourth of the 500 plus projects with a World Bank funded roads component approved over the past decade drew one or more allegations of fraud, corruption, or collusion.[3]

A cursory look at the corrupt practices involved in road making in India trenchantly divulges three discrete components—(a) the existence of unholy nexuses of contracting parties with politicians; (b) Time and Cost over-run; and (c) Poor quality construction. Although other issues, *viz.* dilatory strategies resorted by Government in resolving contending issues involving land acquisition, compensation, resettlement and rehabilitation of displaced persons, legal hurdles, fund crunch etc. impact heavily on the road making exercise, but by and large these issues are amenable. This chapter is an attempt to substantiate the claim that the factors mentioned above have virtually negated the rapid development prospects of the country as the road infrastructure is seen as the engine of growth. A scrutiny is also made to highlight the causes and pattern of corruption and the nexus involved in the implementation of the much hyped rural development scheme of the Government, i.e. Pradhan Mantri Gram Sadak Yojana (PMGSY).

What ails road making in India—the Incongruity between Policies and Performance

The 11th round of annual CMS-India Corruption Study (CMS-ICS, 2017) reveals that nearly 43 per cent of the households across 20 Indian states feel that the level of corruption in public services has increased during the last one year in their respective states. Compared to the scenario about 12 years back, the perception has improved, i.e. a dip of around 30 percentage points is observed (in 2005, 73% of the households had perceived increase in corruption level).[4] The peoples' perception on service delivery by the government speaks volume and indicates quite succinctly the fact that corruption has impacted the lives of commoners to a greater degree. Corruption has become a regular affair in Government offices and transactions. The 2010 Commonwealth Games Scam, 2G Spectrum Allocation Scam, Coalgate Scam, Fodder Scam, Saradha Group Chit Fund Scam, Vyapam Scam of Madhya Pradesh, Rs.1000 crore North Cachar Hills scam of Assam, Rs. 352 crore Mumbai Road Scam are some examples of the prevalence of corruption in the country. The road making scandals are apparently the most *profitable* since at many instances the loot becomes the entire amount of the allocation.[5]

A review of the ills of road infrastructure in India reveals that road making has been the major source of misappropriation of government funds. The Mumbai road scam brought to light the substandard works done by the contractors forcing the BMC (Brihanmumbai Municipal Corporation) to ban seven contractors. The North Cachar Hills scam reveals how the money meant for development purposes are looted by Government officials with support and diktats received from insurgent groups. If one scan through the pages of local vernacular newspapers published from the North-east, one may be able to gauge the plights of commuters as well as Government's apathy towards road making and its development. It is argued that getting a contract approved in road development is all about contacts with the decision-makers. It is even criticised that, sometimes interest groups pressurise the government to allocate inflated funds for road development. These issues echo the incongruous relationship between the policies and the performance of successive governments in road development. Jonathan Lehne and others (2016) explain the existence of political corruption in road construction in India highlighting the indirect evidence that local politicians' power over purportedly neutral bureaucrats is sufficient to coerce them into corruption.[6] It is apparently clear that three significant parameters *viz*. transparency, accountability and quality are compromised in road building exercises in the country.

In order to do away with the malpractices, transparent rules and procedures must be in place. The NHAI (National Highway Authority of India—the nodal agency of execution and monitoring of Highways in India) and the PWDs (Public Works Department—a State Government department to standardise, monitor and execute road building) set standards for building and development of roads. These standards are compromised either by the implementing agency to suit vested interests or by the contractors—whom the contracts are delegated based on open bidding. Off late, a trend of e-tendering has evolved, which is in fact a good step but still it requires stringent adherence to the yardsticks of transparency. Most of the corrupt practices in road making happen primarily because of the maintenance of secrecy and lack of transparency.

Accountability is the other parameter which has long been ignored. In a democracy, people are sovereign entities and the welfare policies directed at them must be in line with the principle of accountability and justice. Accountability does not always mean obeying the contract provider only, but a sense of social commitment must be conceded by the executing parties while implementing projects. This renewed orientation may help reduce corrupt practices to a greater degree not only in road making enterprises but in entire developmental projects.

The third parameter, i.e. quality is the most compromised and ignored parameter. Contractors and firms use low quality material to garner more profit. This has been a tragic reality in Indian road making history. The "all weather roads" become deplorable during rainy seasons. The bad approach roads and poor drainage facilities add to the woes. The Indian roads are impaired by construction hazards; slow pace of works, substandard quality and transparency issues. Unless resolved, these issues may impact the road construction sector for a long period of time.

PMGSY—The Peoples' Road

Rural road building has been a constant and considerate policy of the Government of India since independence. Successive Governments have been seen resorting to the path of rural development through expansion, renovation and development of rural roads. Among the slew of schemes, the PMGSY (Pradhan Mantri Gram Sadak Yojana) has been the most popular and effective as it has transcended rural economy and standard of living to a greater height. The aim of the scheme is so good and appreciable that spirited implementation may potentially reduce the rural-urban gap. Launched on 25th December, 2000 as a fully funded Centrally Sponsored Scheme to provide all weather road connectivity in rural areas of the country, the programme envisages connecting all habitations with a population of 500 persons and above in the plain areas and 250 persons and above in hill States, the tribal and the desert areas.[7]

In the year 2000, an estimated 3,30,000 Indian villages or habitations – out of a total of 8,25,000– were not connected to a road that provided all-weather access (PMGSY, 2004). Their inhabitants were at least partially cut-off from economic opportunities and public services (such as health care and education). The programme has been described as "unprecedented in its scale and scope" (Aggarwal, 2015)[8], with roadwork for over 1,31,292 habitations completed and another 22,696 under construction as of May 2017.[9] A whopping increase in budgetary allocations on road development (Rs. 64,000 crore allotted for the 2017-18 financial year) signifies the fact that roads are central to development. But everything is not right with the PMGSY. Although the policy is laudable but compromise with the quality of works impinge upon the scheme very badly. Since, this is a centrally sponsored scheme and constant monitoring is not possible, hence the contractors manipulate with the quality to extract profit. Although cannot be generalised, this holds true to many PMGSY schemes being implemented.

Unholy Connection—Bountiful Appropriation

"Corruption is rampant in a system run by the uninitiated and the unillumined political leaders because their proven incompetence and immaturity creates infantile disorder. But the moment the greed of the politicians receives overt and covert support of the bureaucrat, corruption becomes institutionalised and rule-based and this inaugurates a regime of sleeze by power-hungry opportunists and upstart adventurists whose lack of rectitude in public life has always been a pet subject of gossip columns, public debate and wall writings" (V.S. Sekhawat, 2005).[10]

The menace of corruption and contractor-politician nexus has badly plagued road infrastructure in the country. Mention may be made of a World Bank funded road project of Andhra Pradesh. The World Bank reported collusion of interests in the *Andhra Pradesh State Highway* constructed during 1997-2004 which took away a whopping 91 million US dollar out from the project. The collusion of interests was detected in two contracts for road widening and strengthening of highways.[11]

Prevalence of rampant and unrestrained sleaze has some connection with the dominant mode of cronyism advanced by the onward march of neo-liberal enchantments. The neo-liberal framework has ushered in an era of market competition resulting in fierce peer rivalry. This rivalry facilitates contacts with politicians and paves the way for favouritism. Favouritism of this kind is the outcome of crony capitalism which has the potential of ruining the very object of the scheme. The unholy alliance is sometimes very deep and rooted.

Delay and Devour—Time and Cost over-run

Another shocking and perturbing phenomena confronting road building in the country is making inordinate delays in carrying out works. Time is money; delay in a certain construction project affects time and thus money, which is the lifeblood of any economy. The timely completion of road construction projects is considered one of the most important factors referring to the project success, as well as the quality and the safety. All around the world many construction projects face one of the biggest construction problems which is the delay, delays differ from a country to another, from a construction project to another and from construction type or cost to another due to every project circumstances.[12] The "Bogibeel rail-*cum*-Road Bridge" being constructed in Assam beats all records in this respect in India and is one of the best example of time and cost over-run. The original target for completion of the multi-crore project was set for April, 2009, but since then, the project has missed several deadlines resulting in massive cost over-run. From an initial outlay of Rs. 1,767 crore, the cost has since jumped to Rs. 3,230.02 crore and railway officials said the final cost is likely to touch. Rs. 5,000 crore.[13] A careful and critical investigation of time and cost over-run reveals quite palpably the fact that although inflated expenditures to a certain degree are justified (due to market fluctuations, loss of materials due to inclement weather etc.) but most of the time it also involves misappropriations. A review of the road building projects in the country exposes the dilatory tactics being employed as a deliberate means to garner profit.

Sacrificing Quality—Poor and Sub-standard Constructions

As has been mentioned above, poor quality construction is the bane of prosperity and development. Existing literatures explicate the fact that poor quality road making not only stresses the economy but also compromise development strategies.[14] A World Bank Policy Study titled "Road Deterioration in Developing Countries: Causes and Remedies" urge the Governments of developing countries to consider the following technical choices that each country should examine in light of its needs and capabilities for ensuring good quality roads—

(a) Gravel roads should be paved only after a complete analysis of the costs, climate, present and future traffic, and reliability of future maintenance.
(b) If the volume of traffic indicates that paving is advisable, the choice of pavement strength should be guided by the likelihood of under-maintenance and excessive axle loads. If the likelihood is high, pavements should be built to the required strength immediately rather than in stages. The cost will, of course, be higher, so that a given budget for new construction will yield fewer kilometres of new road.
(c) Deterioration of paved roads is gradual and barely

perceptible during a long initial phase that lasts up to two-thirds of the pavement's life cycle. Resurfacing and strengthening can thus be deferred somewhat—as long as the roads remain in fair condition and have not entered the critical phase of their life cycle. Longer deferral will cause pavements to break-up. The stage of a paved road in its life cycle must therefore determine the decision to defer resurfacing and strengthening operations.

(d) Traffic is critical for maintenance decisions. If funds are tight, it may be best to keep heavily travelled roads in fair or good condition and to reduce substantially the maintenance of roads with light traffic. Even if funds are not tight, it may make little economic sense to fully restore paved roads that are lightly travelled. If funds are very tight, it may be best to let lightly travelled roads deteriorate further and perform only minimal routine maintenance to keep them usable.[15]

Conclusion

It becomes evidently clear from the above that spread of corruption in road development is a matter of grievous concern. The lack of quality control is although a contributing factor but monetary misappropriations virtually handicap road making endeavours. It is seen that fraud, collusion, and corruption in roads projects wreak enormous damage on developing countries. Roads cost more to build than they should, do not last as long as they ought to, and the corruption proceeds can pollute a nation's political system.[16] These assessments perfectly match Indian experience with road building. In order to do away with the prevailing mode of corruption and to secure Indian roads from further ruin transparency, accountability and quality must be ensured without an iota of compromise. Roads connect cultures and secure livelihoods and hence it is also a responsibility of the citizens to participate in the road building exercise. Active peoples' participation reduces corruption, inaction yields the reverse.

Notes

1. The World Bank, 2011, *Curbing Fraud, Corruption, and Collusion in the Roads Sector,* Washington DC, p. vii, accessed from : http://www.worldbank.org/integrity
2. Government of Gujarat Report No. 2 of the year 2015, Report of the Comptroller and Auditor General of India on Public Sector Undertakings for the year ended 31 March, 2014, p. 87 accessed from : www.cag.gov.in
3. The World Bank, 2011, *Curbing Fraud, Corruption, and Collusion in the Roads Sector,* Washington DC, p. x, accessed from : http://www.worldbank.org/integrity
4. CMS-India Corruption Study 2017 Perception and Experience with Public Services & Snapshot View for 2005-17, CMS Reserach House, New Delhi, 2017, p. 3.
5. Prasanta Mazumadar, *The Indian Express,* "Rs. 1,000 crore North Cachar Hills scam: Three Assam politicians get life imprisonment" accessed from : http://www.newindianexpress.com/nation/2017/may/23/rs-1000-crore-north-cachar-hills-scam-three-assam-politicians-get-life-imprisonment-1608168.html
6. Jonathan Lehne, Jacob N. Shapiro, Oliver Vanden Eynde, 2016, "Building Connections: Political Corruption and Road Construction in India", Paris School of Economics, Working Paper No. 2016-15, *Paris-Jourdan Sciences Economiques,* Paris, p. 25.
7. MoRD, Government of India, accessed from : http://pmgsy.nic.in/Intr_E.pdf
8. Jonathan Lehne, Jacob N. Shapiro, Oliver Vanden Eynde, 2016, "Building connections: Political Corruption and Coad Construction in India", Paris School of Economics, Working Paper No. 2016-15, *Paris-Jourdan Sciences Economiques,* Paris, p. 7.
9. OMMAS (Online Management, Monitoring and Accounting System), http://omms.nic.in/, accessed in May 2017.
10. Vibhuti Singh Shekhawat, "Corruption In India : An Exercise in Perpetuity", *The Indian Journal of Political Science,* Vol. 66, No. 2 (Apr.-June, 2005), pp. 395-404.
11. The World Bank, 2011, *Curbing Fraud, Corruption, and Collusion in the Roads Sector,* Washington DC, p. 6, accessed from : http://www.worldbank.org/integrity
12. A. Sullivan, F.C. Harris (1986), "Delays on large construction projects", *International Journal of Operational Product Manage,* 6 (1) pp. 25–33; cited in Remon F. Aziz and Asmaa A. Abdel-Hakam, "Exploring delay causes of road construction projects in Egypt", published in *Alexandria Engineering Journal,* Volume 55, Issue 2, June 2016, pp. 1515–1539.

13. *The Economic Times*, December 12, 2013, "Bogibeel bridge: Interesting facts about India's longest rail-road bridge".
14. Ebuzoeme, Ogechukwu Divine-Favour, "Evaluating the Significance of Poor Road Design as a Factor of Road Failure: A Study of Onitsha-Enugu Expressway, South-eastern Nigeria", *Civil and Environmental Research*, Vol. 7, No. 2, 2015, p. 180.
15. The World Bank, 1988, *Road Deterioration in Developing Countries Causes and Remedies*, Washington DC., pp. 2-3.
16. The World Bank, 2011, *Curbing Fraud, Corruption, and Collusion in the Roads Sector*, Washington DC, p. 35, accessed from: http://www.worldbank.org/integrity

6

Role of Roads in Maintaining Ahom-Nagas Relation

A Historical Study

I. Imkongakum Ao

Introduction

The first evidence of road development in the Indian subcontinent can be traced back to approximately 2800 BC from the ancient cities of Harappa and Mahenjodaro of the Indus Valley Civilization. Ruling emperors and monarchs of ancient India had constructed roads to connect the cities. Archaeological excavations give us fresh information about road connectivity in ancient India. The Grand Trunk Road was built by Sher Shah Suri in 1549-55 connecting Sonargaon in modern-day Bangladesh with Peshawar in modern-day Pakistan linking several cities from India.

In the case of Ahom-Nagas there is less information about road constructions for their interrelations, the only reference we find is during the reign of *Suklenmung,* the successor of *Suhungmung,* the *Naga-Ali,* which runs through the *Gadhuli Bazar Mauza* from the *Bar Ali* to the Naga Hills, was constructed (Alemchiba, 1970: 34).

Even though there is less mentioned about road connectivity between Ahom-Nagas, even then we cannot even rule out that there were no roads between the two neighbours. Analytical study of Ahom-Nagas points out the course and factors of relationships between the two. Even before Sukapha's invasion of upper Assam, various Nagas were settled in their present habitat. For acute love of freedom, the Nagas led them to a state of perpetual hostility with the powerful neighbours who wanted to bring them under their control. On the other hand, the other reason is said to be "The Shan" chieftains in Burma looked upon the Ahom kings as *Bhai Raja* (Brother King) claiming themselves of descendent from the same royal origin. The Ahoms came to Assam from the Maulung known as Nara Country through the land of the Nagas, and they wanted to maintain their diplomatic relations, with their brother kings—Naras (Shan). But neither the foundation of a new kingdom on the part of the Ahoms in the Brahmaputra Valley, nor their maintenance of good relations with the Naras in the Hukong valley and upper Burma was possible without having a passage through the Patkai pass. The Nagas tried their best to resist the western migration of the Ahom as well as their friendly maintenance with *Bhai Raja,* the King of the Maulung and the Mungkang (Longchar, 2002: 333).

Another reason for relation between the two neighbours was economic necessity in which chilli, pumpkin, cotton, ginger, black pepper, betel vines were commonly produced in the Naga Hills, but by far the most essential of all articles was salt. From Borhat to the interior of the Naga Hills there were several salt springs and a considerable quantity of salt was procured for consumption by the plain men in Assam (Longchar, 2002: 333).

Ahom-Nagas were thus compelled to maintain relationships for all these requirements. Rightly so, the relations grew in all thick and thin situations of both the neighbours. Many historical writings specify on relationship of Ahom-Nagas from political, cultural and economic perspective, but all the so-called relationships were made possible only due to the existence of good communicating roads between the Ahom-Nagas.

Genesis of Ahom-Naga Relations

Ahom-Nagas are neighbours and reside in North Eastern part of India. The study of Ahom-Naga relations reveals sporadic clashes at times but they had cordial and friendly relations most of the time. The two neighbours lived in close proximity for hundreds of years.

The first historical evidence of Ahom-Naga relation can be

traced back to the Ahoms entry to Assam led by their chief Sukapha in 13th century *via* Naga Hills. The historical records explain about the route taken by the Ahom prince in entering the plains of Assam. The Ahoms who entered into North-East Assam from Burma through the Patkai ranges are said to be also Mongoloid people of Tibeto-Burman but of Siamese-Chinese speech, in the eastern most part of the Brahmaputra valley (Alemchiba, 1970: 30).

The first Ahom prince came over to Assam and established a Kingdom was Sukapha. He is said to have left Maulung in A.D. 1215 with a following of eight nobles, and 9,000 women and children. He had with him two elephants and 300 horses. For thirteen years he wandered about the hilly country of the Patkai making occasional raids on Naga villages and in A.D. 1228 he is said to have arrived in Khamjung, from this place he detached his commanders against the king of the Naga whom he ordered them, to destroy. The Nagas put-up a strong resistance as they were not agree to strangers trespassing upon their hills. There were many fierce battles but the Nagas were defeated and perpetrated frightful atrocities on those whom he captured. He caused many of them to be killed and roasted, and compelled their relatives to eat their flesh. This ghastly barbarity created such widespread terror that other Nagas of the neighbourhood all hastened to make their submission. Leaving one of his noble to rule the conquered country, Sukapha proceeded to Dangkaorang, Khamhangpung and Namrup. Then after bridging the Sessa river he ascended the Dihing, finding it unsuitable he preceded downstream and reached Tipam. From Tipam to Mungklang Chekhru (Abhaypur) and again descended towards Brahmaputra to Habung due to flooding in the former place of settlement. But again in 1244 Sukapha continued his journey down the Brahmaputra till he reached the mouth of the *Dikhu*. Thereafter, he went to Ligirigaon and in 1246 he proceeded to Simaluguri, leaving a detachment at Ligirigaon. Finally in 1253, Simaluguri was abandoned in favour of Charaideo where a city was built and furthered the establishment of the Ahom Kingdom in Assam (Gait, 1981: 77).

Soon after the establishment of a new Kingdom Sukapha, the Ahoms developed a new attitude towards their neighbours namely Morans and Borahis, but his fair fame is sullied by the brutal means he adopted to overawe the hostile Nagas of the Patkai hills. The memory of his wanderings along the valley of the Dihing river is still preserved in various local names and traditions. But after the death of Sukapha, his successors realized that a friendly relationship with the Nagas was a precondition for a strong kingdom. So, the Ahom kings adopted the methods of maintaining peace with the Nagas.

Political Relations

As Assam and Nagaland States share territorial boundaries even today. Naga tribes who shares there territorial boundary with Assam are Angami, Ao, Lotha, Konyak, Sumi and PhomNagas, thus, the relationship of Ahom-Nagas cannot be ruled out. Further, it is evident, after the establishment of Ahom Kingdom, the Ahom rulers carried out strategic and diplomatic policies to maintain their hold in the consolidated territory. Thus, the Ahom-Naga relations after the establishment of Ahom kingdom started a new era under the successors of Sukapha.

From the time of Suteupha, the son of Sukapha who succeeded Sukapha to the reign of Susenpha (1268-1488), the accounts are silent so far as the Ahom-Naga relationships are concerned. However, the Nagas were a thorn on the side of the Ahoms, who waged fierce battle on the former occasionally, although they might have not tried seriously to conquer them. In all the Naga raids, the Nagas went down to the plain and plundered the neighbouring villages surrounding the borders. This again points out that even though there was non-existence of well-built roads, the Nagas went through the hilly terrain and jungles making their own path to reach the plains.

In the reign of Susenpha, a son of Suphakpha by a Tipam princess (1439-1488), the Nagas defeated the Ahom forces in an expedition against the Tangsa Konyak Nagas in retaliation to raids committed by them. The king, who led his troops in person, attacked the Nagas, but the Ahoms lost one hundred and forty men in the battle. Susenpha himself fled from the field in a litter, being so overcome with panic that he was purged as he sat there and that the Babrukia Gohain then took command and defeated the Nagas. The Akhampa Nagas then came in with a present of swords in order to make friendship with the Ahoms (Gait, 1981: 78).

During the reign of Suhungmung (1497-1539), the Aitonia Naga revolted against the Ahoms and in 1504, Bar Gohain and the

Burha Gohain were placed in charge of an expedition against the Nagas. The Aitonia Nagas were defeated, and acknowledged the supremacy of the Ahom King, to whom they sent a daughter of their chief and a present of four elephants as a peace offering. They also agreed to pay a yearly tribute of axes, gongs and amber.

In one of the old chronicles we read the history of a childless Borahi, who purchased the son of a Naga of the Banfera clan and brought him up as his own son. The king in appreciation of the work of the Naga boy, made him Barphukan or Viceroy of Guwahati in place of Langi Barphukan. Suhungmung (or Dihingia raja) was also deeply attached to one Karenpa Naga of the Banfera clan and took meals with him in his royal palace. It is narrated that the queen of the late Supimpha (father of Suhungmung) having cast constant glances upon Karengpa excited the wrath of her husband and was made over to the guest. Subsequently a son was born to them who was named Senglung. Suhungmung, on seeing this youth, was struck by his high-bred appearance, and learning that his mother was already pregnant before Supimpha sent her away, he took him into favour and created for him the new appointment of Barpatra Gohain. Which, he made equal to that of the Bar Gohain and Burha Gohain. When the two functionaries objected to the new appointment and refused to give-up for it any of the men under their control. The king however overcame this difficulty by allotting to the Barpatra Gohain, the Barahis, Chutiyas and Morans, who had not been placed under either of the other Gohains. He then called a council of all leading nobles and giving Senglung a seat between the Bar Gohain and the Burha Gohain, publicly invested him with his new appointment and declared his rank to be equal to theirs (Alemchiba, 1970:34).

During the reign of Suklenmung, the successor of Suhungmung, the Naga-Ali (Naga Road), which runs through the Gadhuli Bazar Mauza from the Bar Ali to the Naga Hills, was constructed. The main purpose of Naga-Ali is said to have served two purposes for the Ahoms, firstly, the road was built to have easy access to trade with the Nagas and the other purpose of the road was to provide a route of escape into the Naga Hills in times of raids and invasions from their enemies in the west.

Another interesting incident of Ahom-Naga political relationship can be seen after the death of Chakradhvaj Singh in 1670, in a short space of eleven years, there had been no less than seven kings, not one of whom had died a natural death. During this time no significant event with the Nagas took place. Gadapani ascended the throne in 1681, succeeding Lora raja or Sulikpha and assumed the Ahom name Supatpha and the Hindu name Gadadhar Singh. But it is said before his accession to the throne, Laluk Buraphukon, a claimant to the throne became ambitious and tried to massacre Gadapani. Being fully aware of Laluk's design against him, Prince Gadapani left his ancestral home at Tungkhung, wandered from place to place, disguised as a Naga, as petty trader, and sometimes as an ordinary cultivator. Many places in Assam including the Naga Hills claimed to have been visited by the fugitive prince and stories of his escape from the enemies and of his wanderings, thrilling and colourful, from part of the legendary lore of the Assamese people. While in the Naga Hills, he is said to have saved himself by hiding behind a huge boulder. A Naga lady is said to have protected him once by concealing him inside her skirt. Not only that he is said to have visited the Phom and Ao Naga areas and also was married to Naga ladies of those area, which resulted in giving him two sons Lai and Lechai (Alemchiba, 1970: 36).

Ahom-Nagas at the moments of great stress and strain the Assamese people and their rulers fell back upon "My Hill, So Strong". A king or a prince had to fly to the Naga friends when there was an irresistible foreign invasion or when circumstances at home became unfriendly. On the other hand, there are traditions among some tribes about their chiefs had to be sheltered in the plains when feuds arose among themselves.

Economic Relationship between the Ahom-Nagas

Economically, the Ahom-Naga relation is commendable in history for its uniqueness. The Nagas are a freedom-loving and war like people, who have made head hunting almost a religious institution. This also brought them into conflict with Ahom power and skirmishes often became constant. This necessitated the appointment of some officers like the Kamjangiya Gohain, who was stationed at Khamjang and a number of Naga Katakis were set-up in the bordering areas. The purpose of Kataki was to maintain good relations with the Nagas, who admitted Ahom supremacy and with other tribes remaining independence. It was the custom for the Nagas to leave their spears at their '*Katakis*' house as they come down from the hills on trading expeditions and reclaim them on their return.

Another, feature of economic relation can be seen in a number of agricultural farms (which became known as *Nagakhat*) and fisheries were reserved in the plains for the sole use of the friendly tribes who were called *Bori* (submission) Nagas, as distinguished from the *Abori* (Non-submissive) Nagas. Some Ao and Lotha tribes enjoyed such privilege. Moreover, the Boris was always to be protected against the Aboris. The Bori Nagas carried trade with the Ahom territory, one of the principal centres of this trade being Borhat (Alemchiba, 1970:37).

The Nagas brought down their agricultural produces, rock salt, Naga Daos and different medicinal herbs and they received from the Ahoms the Plainsmen rice, other food staff, medicines, beads, cloths etc. in exchange (Alemchiba, 1970:37).

The *Katakis* were also regarded as special markets for the Nagas. It is said that the *Katakis* served for the Ahom in both diplomacy and business with the Nagas. There were many *Katakis* in different places and their functions were to maintain a channel of communication between the Ahoms and the Nagas as well as to preserve peace and harmony at the Ahom-Naga boundary. All the trading activities between the Ahom-Nagas were carried out by means of barter system and in most of the cases the bordering Nagas visited Assam for all the business related activities.

Cultural Relationship between Ahom-Nagas

If political relations may lead two peoples either to amity or to hostility, cultural relations most generally do promote only friendship and understanding. Political relations change, while the cultural ones persist. This is particularly seen in two things such as language and religion. Religiously, the impact is less noticeable but linguistically the impact of Ahom-Naga relationship is more prominent.

The Naga languages have many features in which they differ from each other, the differences leading to an insurmountable difficulty for each group to understand the other. In the accounts of the British explorers of the last century, we find mention of one group of Nagas bypassing another without a single syllable being exchanged between the two groups. The Nagas have, however, found it possible to bridge this gulf among them through the medium of the Assamese language, which has long ago acquired many Indo-Mongoloid lexical and other features. They have got a basic form of this language of their own, which is remarkable for its simplicity and directness (Alemchiba, 1970:38).

In spite of the varying degrees of remoteness, one fact is sure that the hills and plains people have been living in intimate contact on account of the exigencies of neighbourliness and other indispensible requirements of close proximity. From time immemorial, there used to be regular trade between the Nagas and the people of Assam. Thus, in the course of interaction the Nagas learned the Assamese language, but with different variations in tone and lexical which latter came to be known as Naga-Assamese or *Nagamese.*

Moreover, the close proximity between Ahom-Nagas also have its impact in the Ahoms marring tribal girls and enterprising tribal, who could show their mettle in any deed of valor, bore marks of future eminence were admitted into the Ahomfold and became to all intents and purposes equal in rank and status to the other Ahoms. Some family names of the Ahom bear traces of these infusions, *Naga Patar, Miri Sandikoietc* (Alemchiba, 1970:38).

Military Relationship

The tribals were recruited to the Ahom army in recognition to their muscular strength and capacity for endurance. A British military officer affirmed that, "the Assamese army appears at the time (1660) to have been largely recruited from Nagas and Miris and it is evident that they were quite able to hold their own against the well-trained armies of Hindustan". Soldiers from the tribal areas formed part of the great army of *Lachit Barphukhan.* The General referred to them on one occasion by saying, "Numerous chieftains of the frontiers have joined our ranks. Some of them sought a diversion without consulting us. A detachment has been lost; we have many more still fully prepared for action: King Rudra Singha during his preparations for invading Bengal, assembled at Guwahati an army of our laces which included recruits from the tribal chieftains of the neighbouring regions. In the year 1830, Piali Barphukan attempted to organize a confederacy of the Assam tribes to wrest back the country from the hands of the British; his emissaries scoured the courts of the chiefs of the *Khasis, Jayantias, Bhutias, Nags, Singphos, Khamptis, Miris, Mishmis* and *Manipuris* (Alemchiba, 1970:39).

When the Burmese ruled Assam from 1816-24, many Ahoms ran away from the clutches of the Burmese army and took shelter in the Naga Hills. The Naga provided protection and gave them hospitability until the Burmese were defeated by the British in the Anglo-Burmese war in 1824-26. By *The Treaty of Yandaboo* (1826), the Burmese withdrew from Assam. The Ahom kingdom came under the rule of the British East India Company. Even after the annexation of Assam by the British, the Ahom-Naga relation continued and it was through the Assamese that the Britishers learned about Nagas.

Conclusion

The history of Naga relation with the Ahoms is a blend of hostility and friendliness. The Ahom kings had no intention to conquer the Naga Hills and they never regarded the Nagas as a whole as their subjects. There used to be frequent raids by the eastern Nagas, particularly those living in the low hills near the plains. The Ahom kings punished them by sending expeditions, which often ended in fierce fighting and loss of many lives. A few bordering villages recognized the supremacy of the Ahom Kings and Naga people was maintained through a confederation which was periodically held in the capital of Ahom Kings. It was established mainly for trade purposes of the Nagas in the plains and in return the latter were restrained from committing outrages in the plains.

The relationship was renewed and re-affirmed from time to time and gifts and presents were exchanged as protocol formalities. It was said that the relationship was very strict and delicate in the sense that slight violation of its conditions entailed compelled blockades of 'Doars' for the Nagas until indemnity for its violation was duly paid-up. An understanding was maintained to keep peaceful relationship that in event of any Naga Village committing outrages in the plains, the culprits would be punished by other villages jointly to save 'Doars' from blockades.

References

Achumi, Piketo, 1992, 2012, *British Policy and Administration in Nagaland 1881-1947*, New Delhi: Scholar Publishing House.

Alemchiba, M., 1970, *A Brief Historical Account of Nagaland*, Nagaland: Naga Institute of Culture, Kohima.

Arya, Aditya, 2004, *The Land of The Nagas*, Ahmedabad, India: Mapin Publishing.

Bareh, H., 1970, *Nagaland District Gazetteers*, Government of Nagaland, Nagaland.

Elwin, Verrier, 1969, *The Nagas in the Nineteenth Century*, Oxford University Press, Bombay.

Gait, Edward, 1905, 1926, 1981, *A History of Assam*, Khangendra Narayan Dutta Batuah, Guwahati, Assam, India: L.B.S. Publications.

Mackenzie, Alexander, 1999, *The North-East Frontiers of India*, New Delhi: Mittal Publications.

Mills, J.P., 2003, *The Ao Nagas*, Directorate of Art and Culture, Government of Nagaland, Nagaland.

Mills, J.P., 1980, *The Lotha Nagas*, Directorate of Art and Culture, Government of Nagaland, Nagaland.

Mills, J.P., 1982, *The Rengma Nagas*, Directorate of Art and Culture, Government of Nagaland, Nagaland.

Shimray, R.R., 1985, *Origin and Culture of Nagas*, New Delhi: Samsok Publication.

Section B
Culture and Connectivity

7

Road and Riverine Communication between the Ahoms and the Kacharis

The Kapili and the Dhansiri

Tridib Mahanta and *Bijoy Mohan*

Introduction

The Brahmaputra Valley, in which the Ahoms under Sukapha created a tiny kingdom in the early part of the 13th century, which emerged as the most powerful state occupying the major share of the Valley in later times, has been surrounded by numerous hillocks and mountains where many tribes lived since pre-historic times. The region that we call it today the North-Eastern Region of India comprises all these mountainous tracts and the Brahmaputra Valley. The Ahoms had the major neighbouring kingdoms or countries and tribes such as Bhutan, the Aka (Hrusso), the Dafla (Nishi), the Mising (Hill Miri), the Adi and the Abor on its northern frontier; the Khamti and the Singpho to its east; various Naga tribes, the Kingdom of Manipur, the Kachari, the Jayantia Kingdom, the Khasi and the Garo on its southern side; and the Koch Kingdom in early period and Bengal occupied territory of Goalpara region under Turko-Afghan and Mughal rulers in later times to the west of it. The Kacharis or the Dimasa Kacharis, who once had territorial control over the tract of land in the Valley from the west of Dikhow River to the east of the Kapili River and then pushed back to Doyang-Dhansiri Valley gradually in later period. Since the Ahom State happened to be the biggest player in the entire region, it was obvious for the Ahoms to establish a relation with the neighbouring tribes and communities, and *vice versa*. Hence, opening some line of communication with these tribes and political entities was a necessity for them, and that was possible only by finding some safe, suitable and efficient routes.

Objectives and Scope of the Study

The objectives of this chapter are threefold and they lead to a scope of—(a) making an analysis of physiographical background of the Brahmaputra Valley and its surrounding hilly region in the perspective of the congeniality of finding and establishing routes for communication; and (b) studying comprehensively the Kapili Route and the Dhansiri Route to communicate with the Kacharis and penetrate the Kachari capital and preparing a list of other routes established by the Ahom power in order to reach out to their neighbours.

Methodology

The methodology we have adopted, while dealing with all those issues as stated in the objective, involved collection of evidences from the literary and archaeological sources and their analysis for a logical outcome to be achieved. All these necessitated the division of the work into the following—(a) examining the primary literary sources, more particularly the old Assamese chronicles or *buranjis*, and more particularly the Kachari Buranji, edited by Dr. S. K. Bhuyan; (b) examining the available current literature as secondary sources of information; (c) analying all the information and data through different perspectives or parameters, i.e. inclusiveness, completeness, authenticity or reliability, integration of different source materials, particularly the geographical analysis and Google map analysis etc.; and finally (d) arriving at the conclusion on the basis of the findings.

Physiographical Background of the Brahmaputra Valley and Its Surrounding Hilly Region

The Brahmaputra Valley was the core of the region and at present too from which the Province of Assam has emerged. It is situated between 22°19' and 28°16' north latitude, and 89°42' and 97°12' east longitude as mentioned in the *Gazetteer of Bengal and North-East India.*[1] S. L. Barua has stated that—"The modern state of Assam is but a part of the extensive kingdom of Pragjyotisha or Kamrupa."[2] But, the territorial extent of the land in those days has been variedly mentioned in the early literature. *Kalika Purana,* an *upapurana* compiled in Assam in the 10th century has recorded its extent from the River Karatoya to Dikkarbasini, and it was triangular in shape, 100 *yojanas* in length and 30 *yojanas* in breadth.[3] A 16th century treatise, *Yoginitantra* has stated the boundary of ancient Kamrupa kingdom as the Mountain Kanchanjangha in the north, the confluence of the Brahmaputra and Lakshya rivers in the south, the River Karatoya in the west and the Dikshu river in the east.[4] Hiuen Tsang, the Chinese pilgrim, has noted in the middle of the 7th century that—"On the east this country (the state of Bhaskarbarman) bounded by a line of hills, so that there is no great city to the kingdom."[5] Hiuen Tsang might have meant that the eastern boundary of the state of Bhaskarbarman was bounded by the hills of present Arunachal Pradesh and the Naga hills. After the fall of the Kamrupa kingdom, the extent of the land must have shrinked to a considerable level. As we have already discussed, the Ahom rulers are stated to have tried to establish Sadiya-Karatoya as their east-west frontier, but for the most period of their rule it was from Sadiya to Manah in the east-west direction covering both the banks of the Brahmaputra river, but leaving aside the far southern part of the Dhansiri valley. The Brahmaputra flows through Assam from the east to west ranging over a span of about 900 kilometers at present. Variously named as Luit, Lauhitya, Bar or Bara Luit, the Brahmaputra has been the backbone of Assamese life and culture since early times.[6] The breadth of the river is not uniform throughout its length. The average breadth is 2 kms. The river receives enormous volume of water from its northern and southern tributaries within Assam.The territories of Assam could be broadly divided into two regions—plains and mountainous tracts. While the plain areas are almost covering the state from the west to the east, but broken by some small hillocks in between, the mountainous regions are concentrated on the southern sides starting from Mikir hills or present Karbi Anglong district. In the plains two valleys have emerged as result of crisscrossing of two great rivers—the Brahmaputra entering east Assam from China to the west having finally merged into the Bay of Bengal through present West Bengal and Bangladesh, and the Barak river in the extreme south. Intersecting the main land by the Brahmaputra created three geographical regions—north bank or *uttar kol,* south bank or *dakshin kol* and Majuli, the river island. Both *uttar kol* and *dakshin kol* could be further divided into two parts—Upper Assam and Lower Assam, of which the tract of land from Kaliabar to the west belongs to the former and land beyond Kaliabar to the east lies in the latter.[7]

The Brahmaputra river has supplied a big volume of water by numerous tributaries from the east to the west, flowing from northern and southern mountainous tracts. J. P. Wade, who accompanied Captain Welsh in his Assam expedition in the late 18th century, has given a list of such rivers in his *A Geographical Sketch of Assam* attached with *An Account of Assam.*[8] Of all the tributaries of the Brahmaputra, the Bharali in the north bank and the Dihing are the largest. The crisscrossing of the great river and its tributaries has made the soil of the land alluvial, fertile and suitable for agricultural pursuits. Moreover, all these rivers paved the way for a riverine trafficking in those days. In fact, river navigation or riverine route was considered more convenient than land route. Wade has mentioned that many of the rivers were navigable in all seasons by boats, and a few in the rainy season only.[9] These rivers including the Brahmaputra were thus the integral parts of the socio-economic life of the people of Assam. However, on the other hand, over-flooding of all these rivers during monsoon, caused inundation of vast tract of territories, creating life of the land to pass through immense hardship more particularly for the strangers. Assamese chronicles make mention of all those rivers in connection with travel, flood and warfare. It should be noted that Ahom system of army was largely shaped by the land's topographic conditions.[10]

Stating about the surrounding hilly region of Assam it should be mentioned that there lies a series of mountains or mountainous tracts which led to forming of six territorial entities or presently called Indian states—Arunachal Pradesh, Manipur, Meghalaya, Mizoram, Nagaland and Tripura. Assam is surrounded in the north and east by the hills of Arunachal Pradesh and Bhutan and in the south by Nagaland, Manipur, Meghalaya, Mizoram and Tripura. In the western vicinity of the state there lies the territory of Bengal.

Assamese chronicles and early literary works like the *Kalika Purana* and *Yoginitantra* speak of its hills and dales. *Kamrupar Buranji* has an appendix wherein stated that King Rudra Singha (1696-1714) tried to identify the hillocks around Guwahati with the help of ancient texts.[11] A scholar of modern times, A. K. Bora has recorded in the *Geography of Assam* that the floodplains in the south bank from the eastern proximity down the foothills of Patkai and Naga mountains starts being broken by Chirang Pahar (189 metre high from sea level) in present Nagaon district. He has further mentioned about the other important hillocks of the south bank in westward direction as—Kukurakata Pahar (270 metre), Bura-mayang Pahar (348 metre), Tatimara Pahar (160 metre), Hatisila Pahar (100 metre), Kharghuli Pahar (216 metre), Kamakhya-Nilachal Pahar (293 metre), Dakhla Pahar (138 metre), Surya Pahar (221 metre), Hulokanda Pahar (131 metre), Pancharatna Pahar (313 metre), Paglatek Pahar (200 metre), Moronga Pahar (221 metre) etc. His list includes the important hillocks in the north bank in east-west direction near Tezpur with Bhomoraguri Pahar (140 metre), Singiri Pahar (172 metre), Kurua Pahar (203 metre), Dirgheswari Pahar (329 metre), Agyathuri Pahar (205 metre), Rakhasini Pahar (93 metre), Hatimura Pahar (228 metre), Mahadeo Pahar (110 metre), Chakrasila Pahar (216 metre), Bhairav Pahar (498 metre), Biswakarma Pahar (320 metre), Chandardinga Pahar (245 metre), Dudhnath Pahar (210 metre), Tokrabandha Pahar (258 metre) etc.[12]

The entire region has a damp climate since ancient past. Recording this nature of its climate Hiuen Tsang has noted that—"On the east this country is bounded by a line of hills, so that there is no great city (capital) to the kingdom. Their frontiers, therefore, are contiguous to the barbarians of the south-west (of China). These tribes are, in fact, akin to those of the Man people in their customs. On inquiry I ascertained that after a two months' journey we reach the south-western frontiers of the province of Sz'chuen (*Shuh*). But the mountains and rivers present obstacles, and the pestilential air, the poisonous vapours, the fatal snakes, the destructive vegetation, all these causes of death prevail. On the south-east of this country herds of wild elephants roam about in numbers; therefore, in this district they use them principally in war."[13] The same climatic condition prevailed during the Ahom rule also. According to Shihabuddin Talish, who wrote in the middle of the

17th century, "Assam is a wild and dreadful country, abounding in danger."[14] He further writes, "The climate of the inhabited and lonely places on the banks of the Brahmaputra suits natives and strangers alike. But at a distance from the river, the climate agrees with the natives, while it is rank poison to foreigners. It rains for eight months [in the year]; and even the four months of winter are not [altogether] free from rain............. The air and water of its hills are like the destructive Simoom and deadly poison to natives and strangers alike. Its plains, by reason of their being girt round by hills, tend to breed melancholy and fear."[15] He also states that when Hussain Shah of Bengal invaded Assam he was accompanied by as many as 20,000 infantry soldiers, and many horses and boats, but due to the adverse climatic condition the invaders had to face defeat at the hands of the defending forces and ultimately Hussain Shah was captured and slain.[16] Talish's description is corroborated by another Persian chronicle *Maasir-i-Alamgiri* where it is recorded that the Mughal army sent for Assam campaign during the reign of Mughal emperor Aurangzeb faced immense hardship in the land due to the adverse climatic condition. The chronicle describes their hardship as such—"When the Khan-i-Khanam took-up his residence at Mathurapur for passing the rainy season and water covered the whole land, the Assamese began to act boldly. .... The poisonous air caused a pestilence which carried off vast numbers. This affected the whole land of Assam, and vast crowds of the enemy, too, in the hills went to hell."[17] The adverse climatic condition of the land is also reflected in the account of a Dutch sailor, Glannius who accompanied Mir Zumla in that expedition. He states—"The water became so infected, by the prodigious quantity of dead bodies thrown therein, that several persons perished by that means; wherefore at length, we *boyled* (boiled) the water, before we used it. After three months stay, the water having been continually at that height, that it was impossible to come out of our trenches."[18] He writes about the rainy season of Assam as such—"the season of Rain came; during which the whole country is overflowed, excepting the villages which stand upon hills. This season, which lasted three months, hindered the Emir's designs, the waters keeping him from marching either backwards or forwards....... This weather, and incommodiousness of the place, having near ruined his (Mir Zumla's) whole army."[19] Bernier also gives similar type of description about the rainy season in Assam—"Meantime, the season of the Rains came in sooner than usually : and they being excessive in some parts and overflowing all the Country, except such villages as stand on raised ground, the Emir was much embarrass'd."[20] It is stated in the *Padshah Buranji* that Shayesta Khan, the Governor of the Bengal province of the Mughal empire in the late 17th century tried to make Ram Singha, the general sent for Assam expedition, aware of the situation of Assam.[21]

Passages to and from Neighbouring Territories

Of course, the Ahoms had to maintain relationship with their neighbours particularly the hilly people belonging to several tribes as we have already discussed here, but an equal or more than an equal situation also arose for the latter, and the reason for that was their requirement of mainly food items which were grown in the Ahom country, but not in their territories, and also some other articles. Many of them made inroads into the Ahom territories and plundered the foothill villagers and took away resources from them. Hence, to their advantages, they opened many routes from their regions to the Ahom State. And the Ahoms in their turn had to establish and maintain entry gate or pass called *'duar', 'khat', 'pam'* etc. Discussing it in the context of the Bhutanese people or the Bhutiyas, we may refer to nine *duars* that were for the disposal of the trafficking of the people of this tribe, and they were from east to west, Charduar, Kariapar, Buriguma, Kalling, Gharkola, Boksa, Chapaguri, Chapakhamar and Bijni. Through these routes some volume of trade happened between Assam and Bhutan in those days. There were two more *duars,* Naduar and Chaiduar, which were not meant for the Bhutiyas, rather for the people belonging to the Daffla tribes or called the Nishis. They also did trade with foothill people of Assam, but sometimes plundering of Assamese village people was done by them as well. Similarly, the northern and north-eastern tribes such as the Akas, the Abors or the Adis, the Hill Misings or called the Ghasi Miris or the Parbatia Miris, the Mishimis, the eastern tribes, the Khamtis and the Singphos, and then the Naga tribes in the southern frontier made inroads to the Ahom country from time to time especially when the Ahom villagers or the Ahoms as a whole were found weak in keeping restraint/impediments for them, and roads to the latter's territory were as such opened.

Besides the passages created by the tribes to make inroads to the Ahom country, there were also some roads which were constructed either at the behest of the Ahoms or directly built by the latter which led one to the lands of neighbouring tribes and communities such as the Aka, the Dafla (Nishi), various Naga tribes etc. In so far as the interest of the Ahoms was concerned, the main purposes behind their construction have been believed to be doing trade and commerce, and maintaining political and diplomatic relations with their neighbours. A few of them are mentioned below through which a kind of linkage was established by the Ahoms with these tribes :

> ***Aka Ali :*** It was constructed in the reign of Gadadhar Singha. It ran from Golaghat to Aka hill.[22]
> ***Jabakar Ali :*** It denotes a place called Jabaka near Charaideo where a particular clan of the Naga tribe inhabited. The purpose of construction of the road was to establish commercial relation with the Naga clan. It was constructed during the reign of Pratap Singha.[23]
> ***Kharikatia Ali :*** It was constructed in the reign of Rudra Singha in order to collect wood, stone and other material to be used in various construction works in the state. The road ran from Paniduar Ali to Naga hills *via* Namti.[24]
> ***Naga Ali :*** It ran from the bank of the River Namdang to Naga hill *via* Charing. It was raised in the reign of Garganya Raja.[25]
> ***Namti Ali :*** The road ran from Rangpur to Naga hill *via* Namti. It was constructed in the reign of Lakshmi Singha under the supervision of one Namtial Barbarua.[26]
> ***Taokakar Ali :*** Name derived from a rivulet called Taokak. This road runs from the bank side of Taokak rivulet to Bengenabari, Jabakial Naga-gaon etc. in present Sivasagar district. Pratap Singha erected a fort there.[27]
> ***Teton Ali :*** The name derived from Momai Tamuli Barbarua, who was also nicknamed as Teton Tamuli in those days. The road extended from south bank to north bank starting from Lakhowgarh in the south to Dafla hill in the north.[28]

Besides these roads, there were many other minor roads which also established linkage between the Ahoms and their neighbouring tribes.

How to penetrate the Kachari territories?

Tracing back to the attempts made by the Ahom rulers to open routes to communicate its Kachari neighbours, it has come to light that despite many difficulties and hurdles a few such could be materialised, and the oldest among them was the Marangi Route. We have found that Suhungmung Dihingia Raja, after pushing back and expelling the Kacharis beyond the point of Marangi in the south, constructed a fort there and posted a high power official called Marangi Khowa Gohainin that place to look after the Kachari affairs. Since then both the powers have been recorded in chronicles using a route which connected Marangi (a place near present Golaghat Town) with the Kachari kingdom called the Dhansiri Route. This route was very common and useful for them till the days of Pratap Singha. But, thereafter a new route came up the Kapili Route, which connected Roha (a place near present Nagaon Town) with the Kachari country. The reason might be the fact that the western boundary by that time got extended to at least Sala-Kaliabar region, or may be somewhat beyond that area up to the Kapili Valley. It is also a fact at the same time that, although Suhugmung defeated the Muslim invaders of Bengal and pushed them back to the Karatowa river situated to the west of Goalpara region, not to speak of the tract of land lying as far as the Karatowa river, even the areas beyond Kaliabar were not under the permanent control and the territories beyond Guwahati were not at all under the rule of the Ahoms up to Pratap Singha, who despite waging almost incessant warfare against the Mughals had to sign peace treaty with the latter ceding the territories lying in the west of the Barnadi river in the north and Asurar Ali in the south.

A visit to Kachari Buranji for tracing out the Two Routes : Dhansiri and Kapili

Dhansiri river is seen today flowing from the southern side to merge with the river Brahmaputra crisscrossing the National Highway 37 near Numaligarh. A place named as Naga Choki of

Marangi, presently situated a few kms south of NH 37 happened to be the starting point of the riverine route, Dhansiri as it can be traced out from the texts of Kachari Buranji. The chronicle refers to certain places and/or war camps of the Ahoms on the bank line of Dhansiri towards the Kachari country where the soldiers used to take rest in the meanwhile when they proceeded to that area. We have, in fact, found two accounts about this route in Kachari Buranji. According to first account, the next point was Deopani which has been recorded to be 900 *tar* or *beo* or 3.3 kilometres from Naga Choki, where there was a war camp called as *'koth'*. The subsequent camp was situated at a place called Dilao which was 37,000 *beo* (135 kms) from Deopani. Since Dilao was too far from Deopani, the Ahom forces usually sent some of their men called *'agua'* to collect information about the condition of road and the entire situation up to Dilao. A place called Siloni could be found next, but this account is silent on describing its distance from Dilao. However, it may be assumed that it might be not too far from Dilao since the Ahom soldiers were described to have proceeded to that camp very easily, and so the case was for Dijowa whose distance from either Siloni or Dilao was mentioned in that narrative. A point farther Dijowa was Chowatoli which was stated to be 2,700 *beo* (10 kms) from Siloni. The subsequent station was Dhankhunda which was 6,000 *beo* (22 kms) far from Chowatoli. The two next points were stated to be Baghmora and Tinsukia which were 5,500 *beo* (20 kms) and 5,000 *beo* (18 kms) far from their immediately preceding points. The next camp was stated to be Nomalar Theng which lied at a distance of 4,500 *beo* (16.5 kms) from Tinsukia. A camp close to Nomalar Theng was Lathi which was situated at a very strategic location. Another place near to Nomalar Theng and Lathi was Lenguria in which Kuoiganya Buragohain once established war camp. The next point was Tarang which was 2,500 *beo* (9 kms) far from Lathi. A place situated at a distance of 2,500 *beo* (9 kms) from Tarang was Naderang, and thence Maibong, the Kachari capital. The distance between Naderang and Maibong is, however, not mentioned in this account.[29] The total figure may be around 151 kms.

The figure of 2,45,000 *beo* (896 kms) for the distance between Biswanath and Naga Choki in the second account of Kachari Buranji might be a typographical mistake in this published copy of the chronicle.[30] It, however, suggests that the distance between

Naga Choki and Deopani was 5,200 *beo* (19 kms). The next camp Dilao was situated at the distance of 5,200 *beo* (19 kms) from Deopani. The Kakojan camp was the next which was situated 6,400 *beo* (23.5 kms) far from Dilao. The following point was Tinmuyani camp which lied at a place which was 3,840 *beo* (14 kms) far from Kakojan, and then Dijowa, which was 4,200 *beo* (15.3 kms) far from Tinimuyani. The next point was Mornoi which was situated at the distance of 3,400 *beo* (12.5 kms) from Dijowa. The Baghmara camp was 5,500 *beo* (12.5 kms) far from Mornoi, and then Gerekoni camp was 5,200 *beo* (19 kms) far from Baghmara, and thence the distance from Gerekoni to Namira camp near Nomal Hill, 4,200 *beo* (15.3 kms), from Namira to Lathi camp, 4,700 *beo* (17.2 kms), from Lathi to Torong, 2,000 *tar* (7.3 kms), from Amlakhi (near Torong) to Naderang, 3,060 *tar* (15.3 kms), and from Naderang to Maibong, 2,000 *tar* (7.3 kms). This account also mentions that the total distance from Biswanath to Maibong was 59,700 *tar* (218.3 kms) out of which a length of 35,060 *tar* (128 kms) was in hilly areas and 24,640 *tar* (90.1 kms) in plains.[31] While mentioning the total distance, the starting point is mentioned as Biswanath, and a distance of 218 kms. between Biswanath and Maibong may be accepted after corroborating this data with other evidences, however, by overlooking the mistakes of not properly measuring the length and/or not adopting an appropriate scale of measurement.[32] The total figure may be around 197 kms.

A third account is to be found in Kachari Buranji pertaining to the Dhansiri Route, but it ends at mentioning the distance up to Tarang camp. It tells us the distance from Naga Choki to Deopani, 9,200 *beo* (33.6 kms), from Deopani to Dilao, 5,200 *beo* (19 kms), from Dilao to Kakojan, 6,400 *beo* (23.4 kms), from Kakojan to Tinimuni, 1,840 *beo* (6.7 kms), from Tinimuni to Samaguri camp at Dijowa, 4,200 *beo* (15.4 kms), from Mornoi to Baghmara, 5,500 *beo* (20 kms), from Baghmara to Gerekoni, 5,200 *beo* (19 kms), from Gerekoni to Namira, 4,200 *beo* (15.4 kms), from Namira to Top of Nomal Hill, 800 *beo* (2.9 kms), and then from Nomal Hilltop to Tarang, 2,000 *beo* (7.3 kms).[33]

Kapili river is found to have originated from Meghalaya and flowing on south to north direction to merge into the Brahmaputra river through present Nagaon District. It has created a fertile valley in the region by its water resources, but is causing distress for the people living in the valley by inundating its considerable part during monsoon. It is one of the largest southern tributaries of the Brahmaputara river. The starting point of the Kapili Route to the Kachari country as mentioned in the first account was Roha Choki. The next point was Teteliguri which was 6,600 *beo* (24.1 kms) from Roha. And thence, from Teteliguri to Jamuna, 4,500 *beo* (16.5 kms), from Jamuna to Kotohani, 4,500 *beo* (16.5 kms), from Kotohani to Betoni, 4,500 *beo* (16.5 kms), from Betoni to Demera (Dimarua), 3,000 *beo* (11 kms), from Demera to Gelemu, 4,500 *beo* (16.5 kms), from Gelemu to Nangala, 4,100 *beo* (15 kms) and from Nangala to Naderang, 2,300 *beo* (8.4 kms). The distance between Naderang and Maibong has not been mentioned in this account again.[34] The total figure may around 124.5 kms.

The second account mentions that the distance from Biswanath to Roha Choki was 27,500 *tar* (100.5 kms). The distance between a place named Dibarumukh near Roha Choki and Teteliguri was 6,092 *tar* (22.3 kms). And thence, the distance from Teteliguri to Jamunamukh, 3,540 *tar* (13 kms), from Jamunamukh to Satgaon, 4,470 *tar* (16.3 kms). The distance from Satgaon to Kotoha (Kotohani) is not found in this narrative. And thence, the distance from Kotoha to Deodubi, 2,018 *tar* (7.4 kms), from Deodubi to Dimarua, 3,400 *tar* (12.4 kms), from Dimarua to Gelemu, 4,500 *tar* (16.5 kms), from Gelemu to Langala (Nangala), 4,800 *tar* (17.5 kms), from Langala to Garajuri, 1,480 *tar* (5.4 kms) and from Garajuri to Naorang (Naderang), 4,240 *tar* (15.5 kms). The distance from Naorang to Daiang Hill and then to Landesa Hill have not been mentioned in this account, but it might not be a long distance. From Landesa to Baila Hill it was a distance of 4,151 *tar* (15.2 kms), and then from Baila to Maibong it was a distance of 5,680 *tar* (20.8 kms).[35] The total figure may be around 162.3 kms.

In the subsequent pages of the second account, the following figures are, however, mentioned representing the distances between various places on this Kapili Route. They are—from Roha to Teteliguri, 6,092 *tar* (22.3 kms), from Teteliguri to Jamunamukh, 3,540 *tar* (13 kms), from Jamunamukh to Satgaon-Kotoha 4,470 *tar* (16.3 kms), from Kotoha to Deodubi, 5,018 *tar* (18.3 kms), from Deodubi to Saralpani, 4,080 *tar* (14.9 kms), from Saralpani to Dimarua, 3,400 *tar* (12.4 kms), from Dimarua to Gelemu, 4,500 *tar* (16.5 kms), from Gelemu to Langala (Nangala), 4,800 *tar* (17.5 kms), from Langala to Garajuri, 1,480 *tar* (5.4 kms) and from Garajuri to Naorang (Naderang), 4,240 *tar* (15.5 kms), from Naderang to Baila

Hill, 4,151 *tar* (15.2 kms), and from Baila to Maibong, 680 *tar* (2.5 kms)[36], and total 51,451 *tar* (188.2 kms).[37]

Analysis and Findings

From the analysis of information and data collected from the Kachari Buranji, it appears to one that these information and data are neither figuratively accurate, nor the names of various places/ point/war camp are uniformly spelled. For example, in regards to Dhansiri Route, the first account says the distance between Naga Choki and Deopani was 900 *beo* (3.3 kms), whereas the second account tells us it was something 5,200 *beo* (19 kms) and according to third account it was 9,200 *beo* (33.6 kms); a total mismatch among the figures, and hence it is confusing. Then, the comparison of the mileage between Deopani and Dilao mentioned in the first account with the second and third accounts takes one to a gross confusion, since the first one gives us seemingly an exaggerated figure of 37,000 *beo* (135 kms) for the said distance, while the other two reveal a figure of 5,200 *beo* (19 kms). This type of figurative mismatch can be observed in these narratives while mentioning the distances of other points from their nearest ones. Moreover, the figures for total distance between Naga Choki and Maibong found after calculation of data given in these accounts do not match with each other. Besides, there was disparity too spelling the names of various places or points, such as naming one place as Tinsukia in the first account is nowhere in the second and third, and might be in its place, a place was spelled as Tinmuayni or Tinimuni, and that too beyond the point what has been referred to in the former account. Likewise, the names of the places such as Chowatali, Dhankhunda, Siloni, Lenguria, Amlakhi etc. have not uniformly mentioned in all three accounts.

Figurative mismatch is also found in the two accounts of the Chronicle for the Kapili Route. For example, the distance between Naderang and Maibong is calculated from the narratives of the second account to be at least more than 10,000 *beo* (36.5 kms), whereas the said distance is not mentioned in the first account. Besides, this figure does not match the one mentioned in the second account for the Dhansiri Route, which is stated there as 2,000 *tar* (7.3 kms). Moreover, the total figures for the distance between Roha Choki and Maibong found from calculation of the data given in the narratives of both the accounts do not resemble. Then, pertaining to disparity in place names, the places such as Dibarumukh, Satgaon, Deodubi, Garajuri, Baila, Landesa etc. are not mentioned in the first account, and some places like Nangala (Langala), Kotoha (Kotohani), Naderang (Naorang) etc. are differently spelled in these accounts.

Despite these errors in figures and spellings we can have at the same time many useful information and data in these accounts relating to both Dhansiri and Kapili Routes created by the Ahoms to penetrate the Kachari kingdom. It is found that the Kapili Route was shorter than the Dhansiri Route, and it was perhaps more convenient too than the latter. By analysing the data relating to the distances of various places falling in this part land in present day context and with the help modern gadgets and facilities, these narratives are found to be almost correct. For example, the Google map of Assam take us to finding an aerial length of approximately 120 kms and motorable length, 170 kms between Nagaon Town (close to Roha) and Maibong Town. However, the same between Golaghat Town (close to Naga Choki, Marangi) and Maibong Town are 150 kms and 236 kms respectively. Both these aerial values are somewhat close to those calculated from the accounts of Kachari Buranji. Moreover, names of many places given in these accounts are same today such as Deopani, Siloni (Silonijan), Jamunamukh, Satgaon etc., and also these names corroborate other chronicles and texts.

Conclusion

From gathering and accumulation of information and data and their analysis as mentioned above, we have found clear idea of both Dhansiri and Kapili Routes created by the Ahoms to penetrate the Kachari kingdom, which were also used by the Ahoms to go beyond that land to catch or find out way to other territories like the kingdoms of Manipur and Tripura. There may be mismatch, difference, disparity and all that, but at the end there are many useful information and data which can be used as important sources for making physiographical study of Assam from historical perspective. Moreover, there might be an idea in the minds of both the powers, the Ahoms and the Kacharis, and more particularly the latter about undertaking the recourse of inter-country trade and/or commercial ties between themselves apart from exploring the possibilities of diplomatic relation.

Notes

1. *Gazetteer of Bengal and North-East India,* B. C. Allen, E. A. Gait, C. G. H. Allen & H. F. Howard, reprint, Delhi, 1979, p. 16.
2. S.L. Baruah, 1955, *A Comprehensive History of Assam,* New Delhi, p. 73.
3. *The Kalikapurana : Text, Introduction and Translation in English,* (ed.) B. N. Shastri, in 3 vols., Delhi, 1991, vol. II, ch. 51, *sloka* 76/77, p. 726; E. A. Gait, *A History of Assam,* 4th edn., reprint, Guwahati, 1990, p. 11; S. L. Baruah, *op. cit.,* p. 75.
4. *Yoginitantram,* tr. and revised as *Sanubadam: Joginitantram* by Kalimohan Bhattacharya, 4th edn., Calcutta, 1333 *sal,* pt. I, chapter xi, pp. 131f; *Yoginitantram,* tr. and pub. as *Joginitantra : Sanubadam* by Duttabarua & Company, Guwahati, 1972, pt. I, chapter xi, pp. 147f; E.A. Gait, *A History of Assam, op. cit.,* p. 11.
5. *Si-Yu-Ki* by Hiuen Tsang, translated into English as *Buddhist Records of the Western World* by Samuel Beal, vol. II, London, 1884 (digitized and posted in internet), pp. 198f.
6. S. L. Baruah, *op. cit.,* p. 5; *An Account of Assam,* by J. P. Wade, (ed.) B. Sharma, 2nd impression, Guwahati, 1972, pp. 368f; *Hemkosha* by Hem Chandra Barua, (ed.) D. Barua, 12th edn., Guwahati, 2006, pp. 883, 892; *Chandrakranta Abhidhan,* Jorhat, 1933, pp. 924, 934; *AB,* p. 46.
7. *An Account of Assam,* by J. P. Wade, *op. cit.,* p. 341.
8. Wade's list of the northern rivers were—Sowpura, Khobunkhiri (Suvansiri), Khomederi (Somdiri or Somsiri), Pabo (Pabho), Owah, Gayraylooa (Gerelua), Kasicota, Dikrungh (Dikrang), Pisola (Pichala), Burropani (Barapani), Doorpangh, Dehiri, Seinsa Oujan, Karoi (Kharoi), Seingmora (Sengmara), Maddori (Maduri), Doobia (Dubia), Booroi (Buroi), Bayhalli (Bihali), Burrogown (Bargang), Boorigawn (Burhigang), Gilladari (Ghiladhari), Dikolei (Dikorai), Dunkhiri (Dhansiri North), Bhonroli (Bharali), Monguldie (Mangaldoi), Burhnuddi (Barnadi), Seinsa (Sesa), Borolia, Bhonti, Rowrowah, Sowlkoa (Saulkhowa), Booradia and Manaha (Manah). And the southern rivers were—Dilli (Dilih), Dorika (Darika), Dhekow (Dikhow), Namdangh (Namdang), Jazy (Jaji), Konkiia (Kakila), Dussoi (Dichoi), Dholi, Dilkhiri, Gilladari (Ghiladhari), Doyung (Dayang), Dunkhiri (Dhansiri South), Kaligun (Kalujan), Karzoori, Dhurria, Goloka, Diphulloo (Difalu), Masa (Missa), Hanria, Kopili (Kapili), Bhonrullooa (Bharalua), Gurraloo, Sownra and Kolei (Kalah). In addition to those Wade mentions three more

rivers—Sirilooicheh (Siri Luit), Boodalooicheh (Barluit) and Looicheh Gabroolooicheh (Gabharu-Luit), which originated from the water source or depository called Brahmakunda, from which Wade supposed the Brahmaputra had originated too. *An Account of Assam,* by J. P. Wade, *op. cit.*, pp. 351ff ; S. K. Phukan writes, "All total more than 121 important rivers and tributaries are found in Assam within an area of 78,523 square kilometres." *Hydronymy in Assam,* Guwahati, 2003, p. 67.

9. *An Account of Assam,* by J. P. Wade, *op. cit.*, pp. 351ff.
10. A similar expression is given by Imdad Hussain in his Key-note Address made in the national seminar organised by the Department of History, Arunachal University on 18th-19th November, 2003 where he categorically stated—"the main element of that tradition (the military traditions of the Kamatas and the Koches, and later developed by the Ahoms) were the art of fortification, use of firearms and the methods of riverine warfare", which mostly depended on the topographical condition of the land. See his "Keynote Address" in S. Dutta & B. Tripathy, 2006, edited *Martial Traditions of North-East India,* New Delhi, p. xv.
11. *Kamrupar Buranji,* (ed.), S. K. Bhuyan, 3rd edn., Guwahati, 1987, p. 105ff. The hillocks listed during the rule of the king Rudra Singh were—Manikut (Madhaba of Hajo), Madanachal (Kedar), Gokarna (Kameswar), Garurachal (Mokam or Powa Mecca), Kuberachal (Talaya), Makarechar (Bagata), Bahmaboi (Dadara), Alaboi (Dalibari), Agni (Agiathuti), Chila, Kaljorak, Katapahar, Kharagaruarpur, Nakshatrachal, Girikachal, Natakachal (Natuancha), Chatrasal, Nilkhatap, Sita (Dirgheswari), Manasachal, Saptarshi, Kuberachal, Krouncha (Kurua), Dasasvamedh (Mach Puri-khowa), Manikarneswar (Shahburuj), Amuni, Emuni, Girikachal, Chandra, Indra, Asva, Asvakranta, Godanta, Santadhara, Bayu (Juria), Lohiteswar (Saraighatri Buruj), Lathia, Dambhakuta, Madan, Girikachal, Kanduk, Hatimura, Kamdhenu, Ganesh, Bharadvaj, Gandhamardan, Mutrasila, Ravanboi (Lankesh), Barah, Mashya, Ananta, Rakshash (Nairit), Bagheswari, Govinda, Durga, Gajasardvay, Madhyanak, Chiranjiu, Kapili, Kamdhenu, Nomnanga, Takshakadi, Nilachal, Gandhari, Bhibar, Sarania (Chitrakut), Tridasachal, Itakhuli (Mantrachal), Kalki, Kali, Triva, Bhasmachal and Indrachal.
12. A.K. Bora, "Physiological Background" in A.K. Bhagabati, B.K. Kar, & A.K. Bora edited *Geography of Assam,* 3rd edn., New Delhi, 2007, pp. 22f.
13. *Si-Yu-Ki* by Hiuen Tsang, *op. cit.*, pp. 198f.
14. *Fathiya-i-Ibriya,* Ibn Muhammad Wali Ahmed Shihabuddin Talish, tr. Sir Jadunath Sarkar as "Assam and the Ahoms in 1660", Vol. I, Pt. II, *JBORS,* 1915, p. 182; E.A. Gait, *A History of Assam, op. cit.*, p. 141.

15. *Ibid.*, p. 185; *ibid.*, pp. 141f.
16. *Ibid.*, p. 188; *ibid.*, pp. 144f.
17. *Maasir-i-Alamgiri* by Saqi Mustaid Khan, tr. Sir Jadunath Sarkar, Calcutta, 1947, p. 26.
18. *An Unfortunate Voyage to the Kingdom of Bengal* by Glannius, reproduced with notes as "Mir Zumla's Invasion of Assam" in *Bengal : Past and Present*, vol. XXIX, 1925, p. 24.
19. *Ibid.*, p. 27.
20. *Ibid.*, p. 28.
21. *Padshah Buranji*, (ed.) S. K. Bhuyan, Guwahati, 1935, p. 89.
22. *Asam Buranji Sar* by Kasinath Tamuli Phukan, (ed.) P.C. Choudhury, 4th impression, Guwahati, 1991, p. 40; *Asam Buranji* by Harakanta Barua, Sadar Amin, (ed.) S.K. Bhuyan, Guwahati, 1930, p. 60; E.A. Gait, *A History of Assam*, *op. cit.*, p. 170; H. Barbarua, *Ahomar Din*, Guwahati, 1981, p. 230.
23. *Ibid.*, p. 27; *ibid.*, p. 40.
24. *Asam Buranji Sar* by Kasinath Tamuli Phukan, *op. cit.*, p. 41; *Asam Buranji* by Harakanta Barua, Sadar Amin, *op. cit.*, p. 63; E. A. Gait, *A History of Assam*, *op. cit.*, p. 181; H. Barbarua, *op. cit.*, p. 235.
25. *Ibid.*, p. 20; *ibid.*, p. 29.
26. S. K. Phukan, *op. cit.*, p. 100.
27. *Satsari Asam Buranji* (ed.) S. K. Bhuyan, Guwahati, 1960, p. 26.
28. S. K. Phukan, *op. cit.*, p. 105.
29. Kacahri Buranji (ed.) S. K. Bhuyan, Guwahati, 3rd edn., Guwahati, 1984, p. 71ff.
30. This figure might be 24,500 *beo* (89.6 kms) instead, assuming a zero is less in the original manuscript, and hence a re-examination of original script is required.
31. Kacahri Buranji, *op. cit.*, p. 98.
32. *Ibid.*, pp. 93ff.
33. *Ibid.*, pp. 124.
34. *Ibid.*, pp. 76ff.
35. *Ibid.*, pp. 100ff.
36. This figure comes up might be due to typographical error, in which a digit, 5 before this figure of 680 might have been dropped. A comparison with figures in page no. 105 of the chronicle, Kachari Buranji.
37. Kachari Buranji, *op. cit.*, p. 106.

8

Development of Roadways in Assam-Bengal Frontiers Area and Its Impact on People

Tarun Kr. Halder

Introduction

Assam has been covered by thick forest, mountains, deep valleys, wetlands etc. since early times. Boats, bullock carts, elephants and *palki* were the main medium of communication. During the time of various kingdoms in Assam several roads were constructed but were proved not adequate for modern day's communication. With the coming of the British several changes occurred in the sphere of existing communication system. This was possible because of the emergence of industries, discovery of coal, oil and tea. The British at first put much importance on waterways and subsequently, they put importance on roadways and railways. The development of roadways put a great impact on the socio-economic condition of the people of Assam. The British authority constructed roadways and railways through Assam-Bengal frontiers to reach to the important commercial centres and resource producing regions like tea gardens, valuable forest land, river ports of Assam. It was after independence government has adopted significant steps in terms of roadways and more emphasis has been given on the development of transport and communication. Existing road network was expanded and enlarged so that passengers and goods can be carried from interior areas. This region got significant attention of the government as it is the only region which enables people of Assam to reach to West Bengal and then other parts of India. Development of road communication enables them to communicate easily with their kith and kin, official or institutional contact, and reduced distance and so on. It also put a great socio-economic impact on the people residing in the frontiers area. Development of road extends the path of cultural exchange between Assam and Bengal.

Objectives

The main objective of this chapter is to analyse the development of roadways in Assam-Bengal frontiers area. It also studies the impact of roadways development on the people.

Method and Material

The source materials which are available in the form of book, edited book, articles are used as per historical methodologies. Fieldwork on the frontier area, oral testimonials and own observations have been used as empirical data. Thus, attempt is made to make the paper rational as far as possible on the basis of both literary sources and empirical observations.

Development of Roadways in Assam

The development of roadways in Assam was a gradual process. It is because Assam was covered by deep forest, mountains and deep valleys and wetlands over which making of roads and bridges was not possible at that time. But gradually roadways were developed in Assam. Through roadways it was attempted to connect Assam to the neighbouring regions.[1] Several roads were constructed in medieval period under patronization of several rulers. The Ahom and the Koch rulers contributed a lot towards the development of road transport in medieval Assam. Among these the *Dhodar Ali* and the *Gosain Kamal Ali* were big roads.[2] The British intervened in internal political affairs of Assam during 1792-94. In 1792, British Government of Bengal sent Captain Welsh to sort out the internal crisis of Assam. Assam acquired the attention of British traders and compamy's authority many years before the military intervention of Captain Welsh.[3] After the annexation of Assam through Yandaboo treaty in 1826, the existing road transport pattern of Assam gradually began to change.[4] The British industries especially tea industry is considered responsible for the development of roadways in Assam. The Assam Company claimed that it had built or repaired almost 1280 kilometres public roads and constructed 266 bridges as early as 1945. In 1859, the Company's government started construction of roads which was fit to use bullock carts, elephants to carry goods within short distances. It was in 1865 first attempt was taken to construct a road through the valley of Brahmaputra.[5] As a result, as per recommendation of Company's government Assam Trunk Road was constructed in 1866.[6] The formation of PWD (Public Works Department) in 1868 and the Local Boards in Assam was a great step to extend and develop the road networks of Assam. The purpose of these schemes was to exploit the resources of Assam and to send the same to their own land through Calcutta. The real development of road transport took place in 1926-27 with the creation of a Road Board. The Road Development Committee in 1928 suggested for a structural change of the road transport and called for taking over a major share of the road construction activities. On the recommendations of this committee, the government established a Central Road Committee in 1929. In 1942, Assam Transport Organization was set up which attempted to expand the road communication system of the state in full swing. But the interior areas were not connected with roadways network. Only the important commercial centres and resource producing regions like tea gardens, valuable forest land, railway stations, river ports were linked through road network. It was after independence government has adopted significant step in the sphere of road communication system. More emphasis has been given on the development of transport and communication. Existing road network was expanded and enlarged so that passengers and goods can be carried from interior areas. Within a span of 30 years total 20,321 kilometres roads was constructed or repaired. On this 1383 kms. was National Highways, 2131 kms. State Highways, 16,807 kms. other PWD roads. By 1989-90 this number grown up to National Highways 2033 kilometers, State Highways 2106 kilometers and other PWD roads 25,947 kilometres. Interior places were connected with important commercial centres, towns, ports, railway stations, national highways and state highways so that passengers, resource and goods can be carried from these areas.[7]

Development of Roadways in Assam-Bengal Frontiers

Since the early days intercourse with Bengal has been given much important. Assam had commercial, socio-cultural intercourse with Bengal. Trade with Bengal was mainly carried through river and roadways transport. The main river communication from Kamrupa to Bengal, Bihar and Orissa were the Brahmaputra and Ganges. The Brahmaputra was considered as Highway of river communication. Trade, cultural and religious exchanges also took place between Kamrupa and other Indian states through roadways. There was a road from Magadha to Kamrupa *via* Champa, Kajangala and Paundravardhana. There was fast communication between Nalanda and Kamrupa which hold *via* Bengal.[8] Likewise in medieval period, the main centres of trade in respect of Assam-Bengal frontiers were Hadira or the Assam *chaki*, Goalpara, Jogighopa and Rangamati. The agent of the Assam Government, known as the Duaria Barua, posted at the Assam *choky* situated at the mouth of the Manah River. The administration of the region and adjoining, the Assam *choky* was entrusted to an officer called the Kandhar Barua.[9] The Duaria Barua received the goods of the Assam merchants and exchanged them for Bengal products. They gathered the duties on all exports and imports.[10] The main commodities of export from Assam were stick lac, *munjit* or madder, elephants, tusk, pepper, cotton, mustard seeds (*sarsapa*) and silk and *muga* cloths. Imports consisted mainly of salt, which covered nearly eighty per cent of the total value of imported goods.[11] It was Naranarayan in sixteenth century who built an embankment road a distance of 350 miles from Kooch Behar to Narayanpur under his brother Gosain Kamal specially for military purpose, which was also used by the commoners also for various purposes. This was the only roadway which connected regions of Assam and Bengal at that period.[12] During the colonial period, a road was constructed by British Government along the Brahmaputra Valley from Dhubri to Sadiya. Subsequently, it was connected by steam ferry with the road transports of North Bengal and Goalpara. The British also constructed other roads and railways through this region to reach to the important commercial centres and resource producing regions like tea gardens, valuable forest land, river ports of Assam.[13] It was after independence government has adopted significant step in the sphere of road communication system. More emphasis has been given on the development of transport and communication. Existing road network was expanded and enlarged so that passengers and goods can be carried from interior areas. This region got significant attention of the government as it is the only region which enables people of Assam to reach West Bengal and then other parts of India. There are two National Highways which pass through Assam-Bengal frontiers. These are NH-31 from Guwahati to Barahi *via* Cooch Behar and Siliguri, which covers districts of Kamrup, Nalbari, Barpeta, Kokrajhar, Bongaigaon, Dhubri and NH-27 (earlier known as NH-31.C) Chapaguri to West Bengal *via* Srirampur, which covers Bongaigaon and Kokrajhar districts.[14] Apart from several other state highways were constructed in this region like Beltola-Serfanguri road *via* Lakhiganj, Fakiragram, Shakti Ashram, Jadumani, Dotma, Bautim, Futkibari-Kajigaon road *via* Kokrajhar town, Titaguri, Balajan, Amtoli, Sapatgram-Gossaingaon road *via* Tulsibil, Choutara, Bhawraguri and more importantly Dhubri-Kachugaon road *via* Gauripur then runs through NH-31 up to Sahebganj, Barundanga, Tamarhat and then falls in NH-27 at Srirampur. Further, there are number of state highways which connect either NH-31 or NH-27 (also known as NH-31.C) such as Bhawraguri-Kachugaon Road, Sapatgram-Nayahat Road, Gossaingaon-Tamathat Road, Gossaingaon-Sapkata Road and Tipkai-Suparighat Road *via* Debitola and Kajigaon, Halakura-Kaldoba Road, Fakiragram-Bilasipara Road *via* Raniganj, Bangalipara. Apart from these several sub-ways and other PWD roads are constructed in this region and some are under construction (*Source:* Field Study). Through extensive roadways network interior places are connected with important commercial centres, towns, educational institutions, courts, offices, ports, railway stations, national highways and state highways so that passengers, resource and goods can be carried from these frontiers area (*Source:* Field Study).

Impact on People

The development of roadways in this region has a great impact on the people. It is already stated that the region is mainly inhabited by several groups of peoples like the Kooch-Rajbanshis, the Bodos, the Adibashis, the Bengali Hindus and Muslim people. The region is mainly bordered by a river namely Sonkosh from the Bank of Bhutan up to some distance of Tamarhat, which is main barrier of roadway communication between Assam and Bengal. On Bengal side some distance from border area there is another river named Raidak also seems to be remained a hazard regarding Assam-Bengal roadway communication. Gradually, roadways and railways were constructed through this area. Development of road communication enables the people to communicate easily with their kith and kin, official or institutional contact, and reduced distance and so on. It also put a great socio-economic impact on the people residing in the frontiers area. Development of road extends the path of cultural exchange between Assam and Bengal. Apart from national roadways and Railways several state roadways, other PWD roadways are built through this region which enable them to communicate with one place to other place. It has a great impact on the life of traders. The traders large and small from various corner of this region go to Bengal through cycle, motorbike, personal wheeler, bus and trains as well. They bring various products like potato, onion, pumpkin, cabbages, cauliflower, oil, medicine, traditional home stationery goods in cheaper rate and would sell in the markets of Assam on high profit. The traders of frontiers areas like Srirampur, Tamarhat, Gossaingaon, Kachugaon, Bhawraguri, Tulsibil, Kusumbil, Dotma, Chithila, Sapatgram, Halakura, Motorjhar, Sagolia, Agomani, Sonarhat, Golakganj, Balajan, Sahebganj etc. keeps day-to-day communication with some places of Bengal such as Kumargaon, Chakchaka, Falimari, Atra, Bakshirhat, Tufanganj, Barabisha, Kamakhyaguri, Alipur Duar, Cooch Behar, Falakata, Moinaguri for trading purpose. Most of the products are brought from Barabisha, Kamakhyaguri, Alipur Duar, Kooch Behar, Tufanganj, Samuktola etc. On the other hand, the traders of Bengal frontiers areas also carry various products from Assam such as betel-nut, local betel leaves, fish, grass, pig, vegetables etc. The day-to-day communication is possible because of noteworthy roadways and railway network, availability of motor vehicles, trains etc.

The development of roadways also put a great impact on the life of students. Many students from Bengal are pursuing their study in several colleges of Assam like Chilarai College (Golakganj), Halakura College (Halakura), Gossaingaon College (Gossaingaon) and they are keeping day-to-day communication to their institution which one could not imagine before some years ago. But now it is possible because of development of roadways network. Simultaneously, the students of Assam are also pursuing education in the schools and colleges of Bengal under day-to-day communication. The communication between the Bengali people of Assam and Bengal is very significant. Comparatively, the Bengali people of this region are a little bit more dependent on Bengal regarding the matters of education, language, literature, treatment, employment, marriage, settlement etc. The Kooch-Rajbanshi people of this frontier area are keeping official, organizational, confidential and cultural relation with the same group of people of North Bengal for their common purposes. Likewise, the other communities such as Bodos, Adibashis has been keeping cordial socio-cultural, religious relations with the people of North Bengal. All this is possible because of developed roadways network system in the frontier region. The development of roadways and railways provided marketing facilities, effected movement of agricultural products like jute, rice, timber and agricultural land, import and export of foodstuffs, growth of various small industries, reduced cost of transportation, direct and indirect employment, individual and states income, passengers mobility, non-economic aspect of people, education, culture, population etc. (*Source:* Field Study).

Conclusion

To sum up, it can be said that Assam had commercial, socio-cultural intercourse with Bengal since the early times. Trade with Bengal was mainly carried through river and roadways transport. In medieval period some roads were constructed for various purposes. The development of roadways in Assam was a very gradual process. During the colonial period a road was constructed by British Government along the Brahmaputra Valley. The Company's Government constructed roads and railways through this region to reach the important commercial centres and resource producing regions like tea gardens, valuable forest land, river ports of Assam. It was after independence that government has adopted significant steps in the sphere of road communication system. More emphasis has been given on the development of transport and communication. Existing road network was expanded and enlarged so that passengers and goods can be carried from interior areas. This region got significant attention of the government as it is the only region which enables people of Assam to reach to West Bengal and then other parts of India. Through extensive roadway network interior places are connected with important commercial centres, towns, educational institutions, courts, offices, ports, railway stations, national highways and state highways so that passengers, resource and goods can be carried from these frontiers area. The development of roadways and railways provided marketing facilities, effected agricultural land, import and export of foodstuffs, growth of various small industries, reduced cost of transportation, direct and indirect employment, individual and states income etc.

Notes

1. Goswami, Priyam, *The History of Assam*, pp. 184,186.
2. Medhi, B.B., 1978, *Transport System and Economic Development in Assam*, p. 40.
3. Kalita, R.C., *Assam in the Eighteenth Century*, p. 51.
4. Bhagbati, A.K., Bora, A.K., Kar, B.K. (ed.), *Geography of Assam*, p. 245.
5. Goswami, Priyam, *op. cit.*, p. 187.
6. Medhi, B.B., *op. cit.*, p. 41.
7. Bhagbati, A.K., Bora, A.K., Kar, B.K. (ed.), *op. cit.*, pp. 246, 250.
8. Barua, B.K., *A Cultural History of Assam* (*Early Period*), pp. 111-12.
9. Bhuyan, S.K. (ed.) *Assam Buranji*, pp. 50-51.
10. *Ibid.*, p. 54.
11. Baruah, S.L., *A Comprehensive History of Assam*, p. 442.
12. Gait, Edward, *A History of Assam*, p. 52.
13. Goswami, Priyam, *op. cit.*, p. 187.
14. Bhagbati, A.K., Bora, A.K., Kar, B.K. (ed), *op. cit.*, pp. 246, 250.

Bibliography

Bhagbati, A.K., Bora, A.K., Kar, B.K. (ed.) 2001 (1st Publication). *Geography of Assam*. New Delhi: Rajesh Publication.

Bhuyan, S.K., (ed.) 2010. *Assam Buranji*. Guwahati: Department of Historical and Antiquarian Studies.

Barua, B.K., 2003 (4th edition). *A Cultural History of Assam (Early Period)*. Guwahati: Bina Library.

Baruah, S.L., 2007. *A Comprehensive History of Assam*. New Delhi: M.M. Publishers Pvt. Ltd.

Gait, Edward, 2001 (Reprint Edition). *A History of Assam*. Guwahati: Bina Library.

Goswami, Priyam, 2013. (Reprinted). *The History of Assam*. New Delhi: Orient Blackswan.

Kalita, R.C., 1992. *Assam in the Eighteenth Century*. New Delhi: Omsons Publication.

Medhi, B.B., 1978 (1st Published). *Transport System and Economic Development in Assam*. Guwahati: Publication Board of Assam.

9

Trade and Trade Routes between Assam and Bengal *subas* during the Mughal Period

Tahir H. Ansari

During 13th century Bakhtiyar Khilji established his hold over Bengal and after some time he began his campaign for north-east region of India. He himself took one expedition to conquer this region but, he was defeated and with a great difficulty he was able to save his life. After this, during Delhi Sultanate Muhammad-bin-Tughlaq tried to conquer this land but failed. The Mughals under Emperor Jahangir were able to conquer western Assam valley up to Bar Nadi or Kuch Hajo with the help of Kuch armies in 1612. The Kuch kingdom became the Mughal vassal. Ahom ruled over eastern Assam. The Mughals again tried to re-conquer these regions in 1638 and a treaty was signed between Mughals and Ahoms in the same year and Bar Nadi was fixed the boundary between the two kingdoms. Guwahati came under Mughal control. Although, a number of expeditions were taken to conquer this strange land by governors of Bengal but they could not establish proper hold in this region. Ultimately Mir Jumla, the *nawab* of Bengal invaded over Assam, the eastern frontier region of the Mughal Empire and was able to conquer these regions for some time during 1661-63, which had a great significance in relation to Bengal and Assam states. After this, the Mughals were able to hold their sway at Cooch Behar, Kuch Ḥajo and Guwahati for some time but they were expelled from these regions in 1667. During 1670-80, the Mughals re-conquered Kuch Hajo but finally they lost it in 1681.

The Eastern frontier of this Mughal Empire was constituted by a number of states, both monarchical and tribal which now constitute the north-eastern region. These were Bengal, Kooch Behar, Assam, Manipur, Dimasa, Jaintia, Cachar, Tripura, Khasi, Chittagong, Lushai and the Nagas. Most of them were tribal polities which developed into a feudal state formation. All these state formations developed from a tribal polity into a monarchy and their feudalization was strengthened by gradual sanskritization from across the borders. Of these, some areas like Bengal were integral part of the Mughal Empire, while others like Cooch Behar and Assam were in and out of the empire. Tripura and Cachar were frequently invaded by the Mughal State but, were not formally retained under the Mughals for a long period. Manipur and Jaintia states were overrun by the Mughals whenever the State was short of revenue and withdrew soon without incorporating them in the state. The Khasis, Jaintias, Mughs and Lushais then known as Kukis being on the border of Bengal had an interesting relationship with the Mughal state. They often raided the Mughal territory in Bengal and extracted huge tribute in form of loot of materials and slaves which was their way of asserting their sovereignty. As a result of this interaction even though these large number of states and tribal polities were out of the formal state boundary of the Mughals, they were organically linked to the state's structures and institutions. It became evident when the Mughal state declined. With the weakening of the empire, the erstwhile Mughal officials established an independent state in Bengal. Likewise, other states also declared their 'autonomy' and severed their links with the mother state.

The geographical location of these states on Bengal border sealed their fate. These states by virtue of their location were destined to face intervention from their mighty neighbour – the Mughal State. This intervention was mainly of three kinds: political, economic and cultural. The destinies of these states were bound with the destiny of Mughal State, even though they were not organically a part of the Mughal State. Therefore, when the Mughal State was plagued by multiple crisis and decline, these states had also similar fate. Importantly, it was not just political and economic impact, but also cultural impact, which bound their destinies together. The entire region was an organic whole called Eastern India which included Bengal and the entire north-eastern region.

Mughal India's eastern most provinces were Bengal *suba*. But the central authorities' controls over Bengal have always been frail. Whenever the Mughal State encountered any crisis in its regional units seceded and declared its independence. Bengal was one of the first provinces to do so whenever such a crisis appeared. In fact with the eighteenth century saw the emergence of many successor states. Bengal Nizamate was one of them which became autonomous from the Mughal State until it then passed on to the British. Occasionally from Bengal, incursions into north-eastern areas like the Ahom state, Tripura, Manipur, Dimasa and Cooch Behar were made by the Mughals in an attempt to push the frontier further eastward. But it hardly succeeded fully. Most of these states suffered defeat and temporary occupation by the Mughals to be later reverting to their original status. In fact, a vital research showed that the disintegration of most of the pre-colonial States like that of the Ahoms, Manipur, Dimasa, Tripura and Cooch Behar took place simultaneously with the disintegration of the Bengal in the eighteenth century. And it was not accidental, but due to the organic link between Bengal on the one hand and the states on the other. These states survived the Mughal attack but could not survive the disintegration of Bengal.

The advent of the Mughals in the region had tremendous impact on the society and economy, polity and culture of the north-east India. The brief but effective advent of the Mughals had supplanted certain political and revenue institutions of the Mughals in various states. It generated trade and commerce which linked rest of India with the north-east.

Assam's relationship with neighbouring Bengal up to the middle of the seventeenth century was marked by what may be termed as a close-door approach. This aversion towards foreigners was because of the several waves of Muslim invasions from the west, beginning from the early years of the thirteenth century. In relation to close door policy approach of Ahom, Shahabuddin Talesh has given an interesting picture; he writes "no foreigners ever ruled this country in the past. A stranger never put his foot of domination on this land. The entrance and the exit points are very narrow for the outsiders. Likewise, the inhabitants of this country are shy of a trip to other countries. The *Raja* neither allows foreigners to enter Aasham, nor does he permit his own subjects to go out of his kingdom. Before the conflict that took place between Aashamese and us (Mughals), of which details has already been given in the preface, a group of Aashamese used to come near the border of Guwahati once in a year, with the permission of their *Raja* to carry out business. They brought products like gold, musk, aloe-wood, pepper, *Sazaj* and silk clothes and bartered them for salt, sulphur, brimstone and some other Indian goods imported by the traders of Guwahati."[1] But, despite the close-door approach, there was a considerable amount of commercial interaction between the two regions.

The attitude of the Ahom rulers towards foreigners underwent a change after the final expulsion of the Mughals in 1682 and commercial ties with Bengal were resumed. The setting-up of the Hadira *Chokey* (a *chowky* or *chowki,* also spelt *chowky,* is a trading or police post) (1669) during the temporary cessation of hostilities heralded the change in trade ties between the two regions. Finally, it was during the reign of Rudra Singha that Assam-Bengal trade got a new boost. Rudra Singha abandoned the earlier policy of isolation and showed keen interest in developing trade ties with the neighbouring countries.

In one of his communications to the Mughal *faujdar* in 1713, the Ahom monarch ridiculed the former's idea of establishing trade ties with Assam by sending only a few boats laden with merchandise. He wrote to the Mughal official that regular commercial intercourse would mean sending his merchants (*shah-mahajans*) to Jogighopa and Goalpara and Assamese merchants would proceed to Kandahar *chokey* with their goods.[2] Gradually, traders from Bengal settled at Goalpara with their merchandise while the Assamese merchants stayed at Bangalhat. Assam's exports to Bengal were made-up chiefly of *lac* of which 10,000 mounds were reported to have been exported in the year 1809. Other major items of export were cotton, mustard seed, *muga* silk and ivory. However, the balance of trade was never in Assam's favour primarily because of its total dependence on salt imports from Bengal. For instance, in the year 1809, out of a total import of Rs. 2,28,330 from Bengal, salt imports alone accounted for Rs. 1,92,500.

The major food grain product of Assam was rice from the early period. As Shahabuddin Talesh has observed that in upper Assam peasants were not habitual of buying and selling paddy.[3] While in other parts of Assam, however, husked rice—*chaul* formed an item of buying and selling.[4] The rice was accepted as a popular medium of exchange in Kamrup. It means that even to cross a river by boat, one could offer rice to the boat man in lieu of money.[5] Generally, the poor people used to barter their other domestic products for rice.[6] Epigraphic records testify to the sale of rice mentioning its prices. For example, the Sidheswar Temple Inscription puts the price of rice at eight *ana* (Rs. ½) per 40 *sers* or one *man*, but in no foreign or local sources rice is mentioned as an article of export. It is true of Francis Hamilton's account as well which lists all items of export from Assam to Bengal in the early part of the 19th century.[7]

List of the following items are given.

Export-Import List between Assam and Bengal during 1808-09.

Imports to Assam from Bengal

Salt at 5½ rupees	1,92,500
Fine Pulse	800
Ghiu (1000 *mans*)	1,600
Sugar	1,000
Stone beads	2,000
Coral	1,000
Jewels and Pearls	5,000
Cutlery and glass-ware (European)	500
Spices	1,000
Paints	500
Copper	4,800
Red lead	1,000
English woolens	2,000
Taffetas	2,000
Benares silk cloth, called Kink hap	500
Satin	1,000
Gold and silver cloth	1,000
Shells	100
Muslin	10,000
	Rs. 2,28,300

Export from Assam to Bengal

10,000 *mans* of stick *lac*	35,000
Muga silk, 65 *mans*	11,350
Muga cloth, 75 *mans*	17,500
Manjista or Indian madder	500
Black pepper, 50 *mans*	500
Long pepper, 50 *mans*	300
Cotton with seed, 7000 *mans*	35,000
Ivory	6,500
Bell-metal vessels	1,500
Mustard-seed, 15,000	20,000
Iron hoes	600
Slaves, 100	2000
Thaikhol fruit, 50 *mans*	15
	Rs. 1,39,900

It was only in trade with the hill tribes like the Daflas, Miris, Mishimis, Abors, Nagas, Kacharis and Mikirs that rice was exchanged for their hill products. The *Deodhai Asam Buranji* mentions that in a market at Raha in the Mikir and the Kachari border, custom duties included *inter alia* rice implying that this practice was also applicable to the northern tribes.

We get the reference that Mustard seed was collected by the local traders called *baipari* and were exported to Bengal. In the beginning of the 19th century, Francis Hamilton reports that Assam had exported around 15,000 *mans* of mustard seed to Bengal annually.[8] There were some small local markets called *sariyah-hat*, within the Assam state, where mustard seeds were largely sold.[9] Peppers were used as spices and as medicine. Its production was considerably high and its entire surplus in upper Assam was collected by the *mudoi* traders of Kamrup (who came by boat through the Brahmaputra) and exported it to Bengal.[10] In the Vaishnava biographical works, upper Assam was recorded as a land of *agar* wood, black pepper, gold, silk cloth (mulberry), elephant tusks, decorated wicker, i.e. *japi* etc. Hamilton records that it was through the Assam *chowki* at Goalpara that an amount of 50 *mans* of long pepper and 50 *mans* of black pepper were exported to Bengal per annum.[11]

The other domestic products in Assam were areca-nut and betel-leaves, pulses like *mug* and *mati* and jiggery etc. The areca-nut and betel-leaves formed universal items of trade throughout Assam. *Tarikh-i-Assam* is almost silent about trade in upper Assam, mentions of buying and selling of areca-nut and betel-leaves at the Ahom capital, Garhgaon. The book describes, "The only traders who sit in the bazaar are betel leaf sellers".[12]

From epigraphic records, it is learnt that one *guchi* (20 leaves) of *pan* (vine) usually cost 21 *cowries,* and one *pon* (80 pieces) of *tamol* (betel-nut) 1 *ana.*[13] However, trade in areca-nut and vine which could bring only small amount of profit was confined to small traders and *paik*-peasantry as it did not involve large scale capital investment.

In Kamrup district *matimah* cost 5 *gandas* (4 *cowries* = 1 *ganda*) per *ser, mug* 40 *gandas* per *ser,* and *gur* 1 *ana* for one *ser.*[14] Shahabuddin Talesh tells that Assam faced one of the killer epidemic during 1662 and the prices of essential commodities went-up very high. He records as the price of one seer of vegetable became Rs. 14; likewise one seer of grass pea cost Re. 1. Similarly, the price of one *Tola* of opium went up to one *Ashrafi* and one pipe of tobacco for use in *hookah* came to Rs. 3. One seer of *mung dal* and vetch (*mati*) were used to be sold for Rs. 10 and Re. 1 each per *ser* and one seer salt cost Rs. 30.[15] But that is an unusual price of these items. As due to the epidemic and famine in the most parts of Assam, the prices of the commodities had raised.[16] The other products for the trade were cloths of various qualities and sizes. In Kamrup, cotton towels (*gamocha*) were available in the village markets and the price of a *gamocha* was 4 *pons* of cowries.[17] Production of cotton cloths grew as a profession all over the state during our period. In Kamrup particularly some places like Soalkuchi, Hajo, Lakharpara of Rangia, and North Guwahati, Narayanpur in upper Assam were famous for fine and skilful embroidered fabrics.[18] The weaver class peasants of Kamrup were well known for their products. Some of them used to weave cloth on some other's behalf by taking wage at the rate of one *pon* of *cowri* per cubit.[19] The Kuch king Naranarayan is said to have organised the weavers-guild at Tantikuchi in Barpeta where varieties of cotton cloths were produced.[20] A major portion of the cloths were sold out. Besides cotton, silk cloth—*pat, muga* and *eri* (*endi*) were also produced and a part of these were sold in the markets. Hamilton records the prices of these cloths as follows:[21]

Pat silk	Re. 1 to Rs. 8 (as per size and quality)
***Mejankari* silk**	
Dhuti	Re. 1 to Rs. 4
Mekhela	Re. 1 to Rs. 3
Chadar	Rs. 2½ to Rs. 4
Mosquito curtain	Re. 1 to Rs. 6
Riha	Re. ½ to Rs. 3
Eri	
Riha	4 to 6 *anas*
Cheleng	8 *anas* to Rs. 2
Cotton	
Dhuti	Rs. 10 per piece
(Finest one measuring 12 cubits by 3 cubits)	

According to Francis Hamilton, small dealers used to collect the items from peasant families and "hawk them from house to house, for there were no shop and few markets (for cloth)".[22]

The state had to export some items of value to Bengal to meet the need for salt and certain items of dress and luxury for the upper stratum of the society. The export-import relations of Assam particularly with Bengal are evidenced by all contemporary records. In the beginning of the 19th century, it is reported that silk thread and silk-cloths were exported to Bengal.[23] During the later part of our period imported salt was sold in internal markets in small quantities.

The fruits like banana, oranges, mangoes, pineapples etc. were common articles of sale in the markets.[24] Shahabuddin Talesh mentions that ten oranges were sold for one copper coin in the military camp at Gargaon.[25] He further says that these oranges were so big and juicy that without any exaggeration one can have one full glass of juice from one orange. It was very relishing in taste and pleasing in colour. These oranges were available in abundance.[26] He gives interesting information about gold and says that the gold was sifted out from the sand of Brahmaputra. Around ten to twelve thousand men were engaged in that work. Each of them had to deposit one *tola* of gold annually to the treasury of the Raja. But that gold was considered of sub-standard in quality. One *tola* of gold was sold for Rs. 8-9 only. It was told that gold was found everywhere in the sand of Brahmaputra.[27] Occasionally, some other items like *chira* (flat rice), sugar etc. were also sold in small quantities. Hamilton records that during 1808, 1000 *mans* of *ghee,* and sugar worth Rs. 1,000 *were* imported from Bengal.[28] *Ghee* was mainly used for worship rituals and lightening lamps in the temples. The people who came from outside of Assam used ghee as an edible item. It was produced in small scale in Assam and the people consumed it less.[29] As a consequence, when the Mughal soldiers at Garhgaon had to buy *ghee,* it was Rs. 14 per *ser.*[30] Even the milk was hardly sold in upper Assam and as late as the 19th century there was no milk-man to buy, sell and produce milk and milk items on a professional basis.[31]

Some other products of bamboo-work like *jakai, khaloi, pachi, kharahi, dala, pala, kula* etc. were sometimes sold.[32] Horses, bullocks, cows, goats, domestic and wild birds like ducks, pigeons, *tuni* bird and fishes sometimes formed items of trade.[33] It appears that the horse was very costly in the north-east region and sometimes the cost of one horse was Rs. 70.[34] It also seems that the artisan classes like potter, blacksmith and goldsmith might used to produce some items and sold them to obtain other necessaries of living. We also find the reference of purchase of earthen pot (*malacharu*) by the *paiks* spending two *cowries* for each.[35]

The price of the commodities varied from region to region and depended on the qualities of the products. As the price of a towel (*gamocha*) was 4 *anas* at Kamrup, the same product cost 1½ *ana* in upper Assam.[36] The situation was different in upper Assam. Here the goods were abundant and cheap and trade as mentioned above had not developed much. That is why the traders of Kamrup considered upper Assam as "Paradise for merchants", where all valuable commodities could be procured in good quantities and at considerably low price.[37]

Trade in Assam was not much developed in comparison to the Mughal India. There was no organised money-lending community or system to issue *hundi* or *bil* of exchange and any *bima* or insurance system in Assam which was well developed in Mughal India. The lending and borrowing of money was practiced in a very limited way and it was done at individual level in Assam.[38]

There are various references to overland and river routes to Bengal. The river route began from Goalpara down to the Brahmaputra *via* the Jennai from Jamalpore and then after some distance along the Pabna river, a navigable branch of the Ganges. The first overland route from Bengal to Assam lay through Murshidabad, Dinajpore, Rungpore, Bagwa and Goalpara and it was the line of the Calcutta *dak*. This was the most frequently used overland route. There were also routes connecting Guwahati with Cachar, Sylhet and the Jayantiya kingdom.[39] In case of land routes, there were problems of carriage. The manpower force was required for trade and generally peasants were employed for this purpose. As a matter of fact, at the time of war and harvest the traders did not find numerous labour for carrying goods of large amount. The small traders employed female and male labour to carry baskets of goods on their heads or shoulders.[40] Heavy goods which could not be carried on heads or shoulders then the single load was divided into two and was tied into two ends of a bomboo bar called *dang*, or a flat bamboo piece called *biriya* or *baoka*. The *dang* along with the loads was called *bhar*. When the goods were too heavy to carried on by a single man in *bhar*, it was done by two men by binding the loads in the middle of a strong bamboo-pole, and depending on the weight of the load, even by four men by putting the goods on a square or rectangular size flat bamboo appliance bound to two strong bamboo poles called *sangi* or *changi*.[41] Unlike the *banjaras* in Mughal India who organized the transport of goods and their regular movements through pack-oxen[42], there was no such professional trader-*cum*-carrier class of men in Assam on land routes in our period. Haliram Dhekialphukan in the early part of the 19th century observed that in Assam there was no bullock cart; goods were carried here by individual men on their backs and by boat. This was because; only few Assamese traders could go beyond the border for trade. In lower Assam, however, on special occasions bullocks were used to carry goods.

Transportation of goods to distant places was not only conditioned by lack of good overland routes or of man-power carrier but also by geographical and topographical conditions in Assam. In this way, the distant trade was business of rich people only who could collect labour and take the risk of revering trade. The small traders, therefore, confined their business to their own locality only.

But it was the river route through which most of the commercial traffic passed.[43] The most convenient way of trade means was boat. The Brahmaputra River with its tributaries provided the river way between upper and lower Assam and between Assam and its neighbouring states, chiefly Bengal. It is noticed that there was traffic over the river due to heavy trade and business between Assam and the Cachari Kingdom. The traders used to move from Sibsagar to Namdang using Brahmaputra and Kapili up to Demera within the Cachari Kingdom. This journey was completed within 12 days. The Mughal traders from Bengal carried their trade with Assam in boats through Brahmaputra. John M'Cosh had noticed that many of the boats were "of enormous size and capable of carrying one or two hundred mounds (*mans*), though cut out of the solid trunk of one single tree.... When bulky cargoes, such as cotton, are brought down the country, a common practice is to fasten two canoes together with transverse beams, so that the canoes remain three or four feet apart; the platform is then loaded with cotton or straw, and in this way they admit of carrying a much larger bulk than they could when un-united and with much greater safety".[44]

The trade and commerce over river routes were full of difficulties and danger. Strong current in the river in the summer, wind, storm and even sand slowed down the movement of the boats. Stephen Cacella writes that big trading boats from Gaur, Rajmahal and Patna moved-up and down the Brahmaputra throughout the whole year and it was a very tedious journey taking 25 to 35 days from Goalpara to Calcutta and about 33-45 days upstream.[45] In the same way, journey through the Brahmaputra within Assam proper was also time consuming. However, to avoid danger of river traffic, the traders moved in groups and formed temporary partnership to do their business in distant lands.[46] Earlier the course of the Brahmaputra near Hajo had a sharp curve and it was full of sands which caused much difficulty to the movement of boats. The Kuch king Naranarayan is said to have straightened the course of the river by getting its bottom excavated in about 1564 was made proper for the smooth movement of the boats in the river.[47]

There were some big traders and some small traders. The *sadagar* or *saud* and *mudoi* both were the big traders who carried on trade through river between upper and lower Assam, and also in the neighbouring states. Small traders were called *beparis,* and the female traders were called *poharis.*[48] There were *mudois* who dealt in some specific items like gold, silk-cloth and silk, salt, mustard-seeds, elephant-tusks etc. and sometimes even in areca-nut. The *Katha Guru Charita* speaks of merchants who traded in salt and betel nut.[49] This work classifies big merchants into three categories: (i) those dealing in gold and jewels; (ii) those dealing in different types of silk; and (iii) those who traded in articles like salt and alkaline called *khar*.[50] It seems that specialization or monopoly in trade also grew to certain extent, leading to the emergence of a professional class of people called Maji who took the responsibility of safe carriage of goods by boats.

Conclusion

The growth and development of trade relation between Assam and Bengal led to an important impact on the people of Assam. It was through trade with Bengal that people got acquainted with some of the finest products of India, specially sugar, varieties of silk and fine cotton dresses, perfumes etc. and also salt which were not available in Assam. The growth of trade in Assam provided a good opportunity for the poor peasant and artisans' people to earn for their livelihood by selling some small items they produced. A small and rich class grew in the Assamese society. They became big businessmen who made profit by collecting the local goods in cheaper rate and selling them outside of the state in higher prices. In this way, Assam was having sufficient scope of a profitable trade that attracted the attention of the British East India Company during 18th century. In 1793, the British established a commercial treaty with the king of Assam. The growth of trade and commerce relation between Assam and Bengal not only brought economic prosperity to the people of these states but also established social and cultural relation between them.

Notes

1. Talesh, Shahabuddin (2009), *Tarikh-e-Aasham* or *Fatah-i-Ibriya*, tr. Mazhar Asif, (Guwahati: Department of Historical & Antiquarian Studies), pp. 52-53. This book has given a detailed description about the Mughal invasion and the conquest of Assam under the command of Mir Jumla, the Governor of Bengal during 1661-63. It deals about the Mughals' battle, encounters and skirmishes against Ahom forces and the difficulties faced by former. It also records some other aspects

of Assam, i.e. the geography, the society, people, diseases, fauna and flora etc.
2. Barpujari, H. K. (ed.), *A Comprehensive History of Assam,* vol. III.
3. Talesh, Shahabuddin, *Fatah-i-Ibriya,* p. 61.
4. Lekharu, U.C. (ed.), *Katha Guru Charita,* pp. 469, 478.
5. *Ibid.*, p. 504.
6. *Ibid.*, p. 331. *See* also Thakur, Daityari, *Mahapurusha Sri Sankaradeva Aru Sri Madhavadeva Charita,* v. 1250-51.
7. Hamilton, Francis, *An Account of Assam,* pp. 45-46.
8. *Ibid.*, p. 46
9. *Buranji Vivek Ratna,* II, p. 87.
10. *Katha Guru Charita, op. cit.*, p. 87.
11. Hamilton, Francis, *op. cit.*, p. 46.
12. *Fatah-i-Ibriya, op. cit.*, p. 62.
13. Gogoi, Jahnabi (Nath), *Agrarian System of Medieval Assam,* p. 147.
14. *Ibid.*
15. Talesh, Shahabuddin, *op. cit.*, p. 112.
16. *Ibid.*
17. *Katha Guru Charita, op. cit.*, p. 87.
18. *The Agrarian System of Medieval Assam, op. cit.*, p. 147.
19. *Ibid.*
20. *Ibid.*
21. Hamilton, Francis, *An Account of Assam,* p. 62.
22. *Ibid.*
23. *Ibid.*, p. 45.
24. *A Comprehensive History of Assam, op. cit.*, p. 142.
25. *Fatah-i-Ibriya or Tarikh-i-Assam,* p. 40.
26. *Ibid.*
27. *Ibid.*, p. 51.
28. *An Account of Assam, op. cit.*, p. 45.
29. *A Comprehensive History of Assam, op. cit.*, p.146.
30. Sarkar, J.N., *The Life of Mir Jumla: The General of Aurangzeb,* p. 322.
31. Hamilton, Francis, *An Account of Assam,* p. 63.
32. *Agrarian System of Medieval Assam, op. cit.*, p. 149.
33. *Ibid.*
34. *Kachari Buranji,* p. 46.
35. *Ibid.*, p. 94.
36. *Agrarian System of Medieval Assam,* p. 149.
37. *Ibid.*, p. 156.
38. *Katha Guru Charita, op. cit.*, p. 553.
39. Robinson, W. (1975), *A Descriptive Account of Assam,* Delhi, pp. 245ff.
40. Gogoi, Jahnabi (2002), *Agrarian System of Medieval Assam,* New Delhi: Concept Publishing Company, p. 154.
41. Gogoi, Jahnabi, *op. cit.*, p. 155.

42. Habib, Irfan (1963), *The Agrarian System of Mughal India (1556-1707)*, Mumbai, pp. 66f.
43. Bhuyan, S.K. (1974), *op. cit.*, p. 55-56.
44. M'Cosh, John, (1975), *Topography of Assam*, Delhi, p. 28.
45. Wessels, C., (1924), *The Early Jesuit Travellers in Central Asia (1603-1721)*, The Hague, pp. 128f; letters from Stephen Cacella, a Portuguese traveller who visited lower Assam in 1626-27 are incorporated in this book.
46. Guha, Amalendu, "Medieval North-East India: Polity, Society and Economy 1200-1750", occasional paper 19 of the Centre for Social Science Research, Calcutta, 1978, p. 17.
47. Gogoi, Jahnabi (Nath), *Agrarian System of Medieval Assam*, p. 154.
48. *Agrarian System of Medieval Assam, op. cit.*, p. 155.
49. *Mudois* dealing with salt were called *Lon-mudoi*, and those who traded in betel-nut were known as *Guwa-mudoi* (*Katha Guru Charita*) pp. 65, 94, 553.
50. *Ibid.*, pp. 492f.

10

Road and Communication System Under the Ahom Rule

Himadri Gogoi

Introduction

The creation of the construction of land route in Assam was started in the 17th-18th century during the rule of the Ahoms who first introduced modern roads including the National Highways. At that time these were called as *ali* or *garh*. Besides the roads they constructed bridges. The stone bridge of Namdang river on N.H. 37 in Sibsagar district is still prevailed. By observing all these we can surely assume about the energetic labour of the Ahoms regarding the construction of road and communication system in Assam. But it is to be noted that means of transport did not grow much than the construction of roads. Thus to maintain distanced social relationship among the people of Assam became problematic.

The concept of transport and communication system emerged in the ancient period. Yet this system was not advanced as compared to the transport and communication system of the medieval period under the leadership of the Ahom rulers. They emphasised more focus on the construction of roads. That is why our main objective is to highlight the various types of the construction of roads which occupied enormous importance regarding the history of Assam in medieval period[1] (Sen, Debnath, 1978: 157). There were several land routes on the western border.

One of these covers from Cooch Behar through Rangamatty to the western limits of Assam. This route was existed when the Muhammedans first penetrated to Assam. At that time, the entire communication system of Assam was river- borne transport. There were trade routes of Assam with the neighbouring states and also with some other parts of India. There was trade route from Sadiya (the extreme east) to Guwahati, Koch-Kamrup, Mughal India. There were also connecting route from Guwahati to Sylhet, Cachari kingdom and Jayantiya kingdom. The means of transport were—elephants, horses, different types of boats, *changi, dola, khatola* etc. During the reign of the Ahom, Assam became famous among the few states of India because they ousted the invading forces like—Moghul, Afghan and Burmese etc. They were also successful in confronting the frequent attack of the west and also periodic tribesman of the northern, southern and eastern borders. By observing the strategic location of Guwahati, it was selected as the ideal place to check the Mughal inroads. The Mughals had to travel upstream *via* before reaching the Ahom's capital in upper Assam. Thus, the mountains of the neighbourhood of Guwahati were fortified and linked with each other by ramparts erected in the plains. On the other hand, in the east, the hill passes through Patkai range connected Assam with Burma. Another route called *bar ali,* connected upper Assam with the Naga hill. The earliest note on geography of Assam is found in Ptolemy's account—Garint has mentioned the important places of the east, south and south-west by utilising Ptolemy's work, these were close to the frontiers of modern Assam. After the decline of the Pragjyotishpura in the 13th century, Ahoms established their kingdom in Charaideo (upper Assam). From this time Assam became prominent regarding route geography. Ptolemy mentioned one of the important places of upper Assam called *Sadiya.*

Objectives

(i) To look into the policies of public work laid by the Ahom rulers.
(ii) To analyse the modern way of construction of roads and bridges under the Ahom rulers which still existing.
(iii) To highlight the dignity and hard labour of the Ahoms while analysing about the construction of roads.

Sources and Methodology: The proposed study has conducted mainly secondary data collection to fulfil the research objectives. Secondary sources like printed books, articles, magazines etc. have been collected from various library sources. For analysis of the qualitative information, empirical method has been utilised.

Review of Literature

In the book, *The Comprehensive History of Assam,* Barpujari mentioned about the different road, i.e. constructed by the Ahoms which linked several hills states with Assam. He also analysed about the various reasons behind the construction of these roads. But he has given very little information about this field.

Jahnabi Gogoi's (Nath) "Means of Transport in Assam during the Ahom Rule": Gogoi discussed here about the various trade routes along with the different roads constructed by the Ahom rulers. She also stated about the various means of transport and communication system during the Ahom rule. On the other hand, she specially focused here about the riverine trade system. Very limited information has been given with regards to the construction of roads. But she emphasised the contribution of the Ahoms towards transport and communication system. Gogoi has mainly focused on the various means of transport and communication system under the Ahom rule.

Debnath Sen's "Road Communication in Ancient and Medieval Time": He described a detailed account about the road and communication system from the ancient time to the medieval period. Here he stated that though the Ahoms introduced modern way of the construction of roads in medieval period, yet in ancient time, the concept of the road and communication system was prevalent actively. He elaborated about the construction of different roads under the Ahoms.

S.L. Barua's *"A Comprehensive History of Assam"*: Barua discussed about the different ruling skill of Ahom rulers. She has given a clear depiction about the various phases of Ahom-Mughal conflict. In this she has stated about the fortification of Assam to protect it from the attack of the Mughals. But she has not given a direct illustration about the road and communication system under the Ahom rulers.

Santanu Kaushik Barua's *"Assam Abhidhan"*: He described

about thousand facts about the history of Assam under the Ahom rulers. He has given a detailed account about the reign of important Ahom rulers. In this he discussed clearly about the construction of roads. Here, he also mentioned about the Ahom architecture.

Priyam Goswami's *"The history of Assam—From Yandaboo to Partition, 1826-1947"*: Goswami described about the polity, society and economy of Assam during the colonial period. She has provided research based information about the topics. On the other hand, she also analysed about the modern way of transport and communication system in Assam.

"Technology in the Ahom state: Non-agrarian": In this book, the author has given descriptions on the various policies of the Ahom rulers while consolidating the public works of Assam. A clear idea about military and civil constructions are mentioned which is an important part of the Ahom rule. In the discussion about the civil construction ramparts, buildings, roads, bridges, gars are included. The various ways of the construction of roads by the Ahom rulers are also discussed in a significant way.

Construction of Roads under the Ahom Rule

Among the different strategies of the Ahom rulers, one of the most important among these is the civil construction. They took effective policies of public work. Besides the other fields, they exposed their distinctive characteristics and attention towards the civil construction. If we observe the various civil constructions laid by the Ahoms, we can get the idea that they vested so much of hard labour and dignity while constructing the roads. The civil construction includes the construction of roads, bridges, ramparts, buildings both religious and secular, embankments and tanks etc. These constructions reflect how these Ahom rulers controlled and maintained their policies for the proper functioning of their state.

This study put specialisation on the construction of road under the Ahom rulers. It is to be mentioned here that roads were constructed in such a way that they survive till today. Changrung Phukanarburanji noted about all these public works in detail. But copy of this book is not yet recovered. Ahom introduced the *Paik* system, through which the public works and personal works had been done by the adults between an age of fifteen and fifty during peaceful days. The *paiks* were forced to provide state service as it was compulsory for them. They were involved with the construction of roads, ramparts, bridges, buildings, *maidams* etc. As there was found availability of manpower, huge public works were done vehemently. A large number of roads were constructed during that time[2] (Handique, 1992).

Under the reign of Chuklegmung Dihingiya raja (1549–52 A.D.) a full-fledged Public Works Department (PWD) grew-up. Under his military and civil works, the construction of trunk roads, embankments and bridges included.[3] (Barua, 1985:69) After the construction of this trade relation among the people of Assam and also outside Assam developed. During his period the Garhgaon tank was excavated; the Garhgaon town and a number of roads. It is to be noted here that the most significant road in medieval Assam is *Gohain Kamal Ali* which was built by Kochking Naranarayan under the supervision of his brother Gohain Kamal in 1563 A.D. to invade Assam starting from Cooch Behar to Narayanpur. Swargadeo Pratap Singha (1603-1641 A.D.) was the first Ahom ruler who improved the communication system in Assam by constructing roads. The civil constructions laid by Pratap Singha, i.e. the construction of roads, embankments etc. visualised his efficiency in administrative works. For e.g. the old trunk road, i.e. *Bar ali* was constructed by him from Sivsagar to Dibrugarh district. *Bar ali* means 'big road'. This *ali* had connection with Garhgaon. He constructed this road from Gajapur town to Garhgaon along the left bank of the River Brahmaputra. He also built various masonry bridges, i.e. Singhaduar (lion gate) which was in Mechagar. Later, this road was extended to the Numali Garhin the west through Dergaon temple. He also constructed garhsto check the advance of the Mughals in Assam. A national highway of Assam, i.e. *Ladoigarh ali* in Jorhat subdivision was constructed by the same ruler. He also built a stone bridge over the Darika river and constructed many bamboo bridges. Thus, his contribution is creditable as he is successful in defending Assam from the advent of the Mughals. After him, Surampha or Bhagaraja (1641 44 A.D.) built *Salaguri Ali* in Garhgaon. The *Cheoni Ali* of JayadhvajSingha in Sivsagar, a road from Ali Kekuri to Namdang and the tank at Bhatiyapara were constructed in his time. After him Chakradhvaj Singha (1663-1669 A.D.) got the throne. He also took interest in construction of roads and bridges. Another ruler from Tungkhungiya dynasty— Gadadhar Singha (1681-1714 A.D.)

also took interest in the construction of roads. The *Dhudar ali* from Golaghat to Jaypur was constructed by him. It was constructed by the Dhodes (i.e. idle men). It is 115 miles in length. It is running almost parallel to the Brahmaputra from the east of the Dhansiri in Golaghat to Jaipur in the Dibrugarh district. On the other hand, *Aka Ali* and two more stone bridges were built by him. His elder son Rudra Singha (1696–1714 A.D.) is also an important figure of Assam history because of his devotion in multifarious activities like construction of roads, digging of tanks and surveying of lands for e.g. *Kharikatiya ali* in Golaghat district. Another example of this the *Garh ali,* the old military road for horses meeting the trunk road at Dergaon and a *Raja Ali,* a highway from Raha to Kaliabar. His younger brother Siva Singha (1714 – 44 A.D.) is famous in Assam history because he constructed an embankment for the protection of future inroads from the Daflas Assam enjoyed unbroken peace in his reign. During Gaurinath Singha's reign, various roads were constructed with historical names like – *Sewani ali, Na ali, Kamarbandha ali* running out of the town (Garhgaon) and the name Hatigarh refers to a garh, i.e. erected by the ruler for safeguarding the elephants which were vital for communication and warfare.[4] (Gait, 1992:171). The *Cheroni ali* was another road, which was constructed in the Ahom period. This road was 30 miles running from near *Bar Ali* in Gaurisagar to Jorhat. There are also roads like *Bakata ali, Rangalugarh, Bhomoraguri ali, Basa* road from Jorhat to Basa, *Kamar bandha* road, *Mohbondha* road, *Dhai ali, Dubari ali, Meteka ali, Aka ali, Bar Phukanor ali, Mori ali* etc.[5] (Barua, 2000:44-51).

Most importantly these roads were well made and generally constructed to a level that would not be subjected to the inundation of rivers. In fact most of the roads are existed till today.

During the Ahom period these roads provided important contributions like—

Ahoms wanted a proper connectivity of their kingdom.

1. To increase their administrative efficiency.
2. They maintained a good defense policy.
3. They wanted to develop their trade route inside Assam and also outside Assam.[6] (Barua, 1933:211).

Apart from these roads another important feature of Ahom

architecture was the bridge construction which is an important part of civil construction as we have mentioned it earlier. Chakradhvaj Singha's Minister Lachit Barphukan constructed the historic bridge of Saraighat to defend Guwahati from the Mughal invasion during Ahom-Mughal conflict. Namdang stone bridge was the best example of the civil construction laid by Ahoms. This bridge was constructed by the Ahom ruler Rudra Singha. Even Tabaquat-I-Nasiri also reveals how the forces of Bakhtiyar were entrapped by the king of Kamrupa by removing the flat stones of the platform and making the bridge impassable. Following which most of his troops were annihilated. This stone bridge was existed till being destroyed by the earthquake of 1897. Most interestingly one picture of this bridge by Captain Dalton was published by Hanay in the Asiatic Society for the year 1851. There he mentioned that this bridge, a remnant of ancient times in Kamrupas situated about 8 miles north-west of Guwahati, on the high alley which, no doubt formed at one time the principal line of land communication with ancient Guwahati or popularly known as Pragjyotisha and western Kamrup, the structure is solid. He has also given some description about the structure of the bridge that was solid, masonry, built without lime or mortar, of the same massive and enduring material (gneiss and granite) found in the nearby hills of Guwahati.[7] (Neog Maheswar, 1974:141). The structure is like a platform with a slight curve 140 feet long and 8 feet in breadth composed of slabs of stone, six feet nine inches long and ten inches thick, numbering five in the whole breadth, resting on an understructure of sixteen pillars, three in a row, equally divided by three large solid buttresses; with a half buttress projecting from a circular mass of masonry forming abutments at each end of the road, there being, in the whole, 21 passages for the water.

Therefore, it can be said that the tradition of masonry bridges of Assam even predates the Ahoms. On the other hand, another stone bridge was constructed during the period of Pratap Singha (1603-41) over the river of *Dorik* in Sivasagar. According to some chronicles, three Neogs were engaged in the making of the bridge. There are also some references to the fact that during the reign of Ahom king Sudaipha (1677-79), a bridge was constructed while Gadadhar Singha (1681-1696) is supposed to have erected two stone bridges at *Jijikhona* between Charideo and Garhgaon and the other Dijoikhona.[8] (Rajkumar Sarbananda, 1985:357). These instances clearly point to fact that knowledge of stone bridges pre dated the stone bride of Namdang.[9] (Bhuyan, 1992:63).

Gradually, during the Chandrakanta Singha's reign Ahom kingdom declined. Finally in Purandar Singha's reign because of the increase of the Burmese attack, British involved in this and thus Anglo, Burmese war occurred in the year 1824 and according to the Treaty of Yandaboo in 1826, British annexed Assam.

Conclusion

From the above discussion we can undoubtedly say that under the Ahom rule, necessary infrastructure had been taken in the important places of Assam. Although, they did not fulfil all the characteristics of an urban area, yet their contribution cannot be ignored. There was a glorious journey of road history in Assam but the development of roads in Assam was a very slow process. Yet, the mountainous terrain and heavy rainfall, the cost of bridging on the numerous rivers made the road construction very difficult. Although, it is also fact that the innumerable waterways of the Brahmaputra made Assam less dependent on roads than other place. Most importantly it is also fact that the road development of Assam was also neglected by the Government even in the post-colonial period.

Most importantly in the colonial period, British Government developed some roads inside the gardens and some of the roads are still existed. Even in the colonial period water transport was very popular. The British Government usually used the water transportation. The British Government exported the tea and some other raw materials like coal and timber from Assam through the mighty River Brahmaputra. British had connected this waterway from Assam to Calcutta and later they connected to London.

When the British officers like Jenkins and Hopkins entered Assam, suggested to construct a road from Cherrapunjee through Nongkhlaw to Guwahati and Sylhet and also a different road from Cherrapunjee to Bishwanath. It was not until 1865 the initiative was taken to construct a road through the length of the Brahmaputra. The road ran along the south bank of the Brahmaputra from Sadiya to Dhubri. During that time there was also a steam ferry with the road system of Goalpara and Northern Bengal. Guwahati was joined by a road to Shillong which was further joined *via* Cherrapunjee, Theraighat and Companyganj with Sylhet and Cachar. On the other hand, from Cacher a bridal path led to Manipur. And most importantly another cart road was constructed from Manipur Passed through Kohima, Dimapur and Golaghat which finally led to the Brahmaputra. Apart from these major roads another roads which were constructed during the period of Ahoms were also reconstructed by the British. For example Dhudar Ali was connected to Tura to Guwahati.

During the colonial period, a great change came to the road history of Assam. Various scholars also argue about the contribution of British on road construction because British used these roads for the purpose of –

1. Security point of view.
2. Exported various natural resources like—Tea, Coal and Timber etc.
3. Maintain a good administrative system.

Today, it is clear that road development in Assam continued to be neglected for a long time. Even as late as 1939, only 14 per cent of roads were metalled. But road communication is very essential for development of a country. Therefore, to develop this region government has to emphasize on the road construction which will lead Assam towards a new enlightenment.

Notes

1. Debnath Sen, 1978, "Road Construction in Ancient and Medieval Assam", *NEIHA*, VIII, Nagaon, p. 157.
2. B.C. Handique (ed.), 1992, *Changrung Phukanar Buranji*, Bhimkanta Barua (ed.), Dibrugarh.
3. G.C. Baruah, 1985, *Ahom Buranji*, Guwahati, reprint, p. 69.
4. Edward Gait, 1992, *A History of Assam*, Guwahati, 5th edition, p. 171.
5. Barua (ed.), 2000, *Buranjiye Porosa Rangpur*, Sibsagar, pp. 44-51.
6. K.L. Barua, 1933, *Early History of Kamrupa*, Shillong, p. 211.
7. Maheswar Neog, 1974, *Prachya Sasanawali*, Guwahati, 1st edition, p. 141.
8. Sarbananda, Rajkumar, 1985, *Itihasesuwarasokha ta bosor*, Jorhat, p. 357.
9. S.K. Bhuyan, 1992, *Atan Buragohain and His Times*, Guwahati, 2nd edition, April, p. 63.

Bibliography

Barua, S.L. (1997), *A Comprehensive History of Assam*, Guwahati: Munshiram Manoharlal Publishers Private Limited.

Barpujari, H.K. (2007), *The Comprehensive History of Assam*, Guwahati: Publication Board of Assam.

Baruah, G.C. (1985), *Ahomburanji*, Guwahati.

Barua, K.L. (1933), *Early History of Kamrupa*, Shillong.

Bhuyan, S.K. (1994), *Lachit Barphukan and His Times*, Guwahati.

—(1992), *Atan Buragohain and His Times*, Guwahati.

Gait, E.A. (1997), *A History of Assam*, Khagendra Narayan Baruah, Guwahati.

Goswami, Priyam, (2013), *The History of Assam – From Yandaboo to Partition*, 1826-1947, Orient Blackswam Private Limited, New Delhi.

Gogoi, Jahnabi (2016), *Historical Geography of Medieval Assam*, Guwahati: Eastern Book House.

Medhi, S.B. (1978), *Transport and Economic Development in Assam*, Guwahati: Publication Board of Assam.

Neog Maheswar, (1974), *Prachya Sasanawali*, Guwahati.

Rajkumar Sarbananda, (1980), *Itihasesuwarasokha ta bosor*, Jorhat.

Sen, Debnath, (1978), *Road System in Ancient and Medieval Assam*, Nagaon: NEIHA.

Contribution to Books

Barua, (ed.), (2000), *Buranjiyeporosa Rangpur*, Sibsagar.

Gogoi, Jahnabi, (1998), *Means of Transport in Assam during Ahom Rule*, Kohima: NEIHA.

Handique, B.C., Dimbeswar Barua (ed.) (1992), *Changrung Phukanar Buranji*, Dibrugarh.

11

Colonial Road Development Policies in Assam: Its Consequence and Significance

Habib Fazlul Basid

In the Sixteenth and Seventeenth centuries many good roads throughout the villages in the Upper Assam were constructed by the Ahom rulers. It was from the time of Ahom rule that Assam had trade ties with China and it occupied a central place in the field of trade and communication link between Gangetic valley and southern China through Assam and northern Myanmar (Lahiri, 1991: 149). But the end of Ahom rule and the Assam's annexation to the British political control marked the introduction of a new commercial and administrative policy into the province. Expedition of Captain Welsh into Assam during the Ahom king Gaurinath Singha in 1792 and the subsequent commercial treaty that he signed, marked the beginning of the British commercial interest in Assam (Kar, 1997:ix). On 28 February, 1793, an Anglo-Ahom Commercial Treaty was signed with the prime object of establishing a reciprocal and entire liberty of commerce between the subjects of Assam and Bengal for all singular goods (Bhuyan, 1949: 340, 342). But after the fall of the Ahom power the roads built by the Ahom rulers became useless due to the lack of maintenance by the British Government. However, during British rule as in case of waterways, the development of tea industry is greatly responsible for the development of roadways in Assam (Medhi, 1978:40).

Road development in Assam had been a very slow process during the British rule. The ranges of mountains within the province, lack of construction materials, scarcity of labourers and the lack of Government efforts delayed the development of road transportation in Assam (Goswami, 1999: 149). However, the importance of the development of road communication in Assam gained momentum as early as in 1827. In 1827, Jones surveyed the Sylhet-Guwahati Road *via* Cherrapunji and in the next year the road from Jayantiapur to Roha was explored. The Cachar-Manipur road, Banskandy to Bishenpur, was completed under Captain Gurthric towards the close of the 1842, but it could not be extended on account of extreme difficulties of the terrain and outlays involved in the project.

As early as 1833, Captain Jenkins surveyed the North-East Frontier and emphasised the need for gradual development and the restoration of the old causeways which were very essential as embankment for protection of cultivation as well as for commercial as well as military communication. He further wanted to have a continuous line of communication on the Southern bank of the Brahmaputra from Guwahati to Bishwanath, then headquarters of Assam Light Infantry, through Nowgong and Kaliabar. He further suggested to the Governor-General in Council for the appointment of a supervisor permanently in Assam to look after the development of the transport network. In June 1836, accordingly, J. Martin was made the supervisor and secretary to the Public Works Department in Assam.

Jenkins soon realized that the development of province depend by and large on the improvement of communication system which was only possible by constructing some new roads in Assam. He wanted to utilize a minimum percentage of revenue for the repair and construction of roads in Assam. When Mill visited Assam in 1953, carts and carriages were unknown and the roads were very few and bad, there were no roads at all in Sylhet and Cachar (Gait, 2008: 398). The viewpointof Jenkins was supported by A.J. Mills and he recommended two trunk roads, one leading from Bengal running alone the South bank of Brahmaputra by Goalpara, Guwahati, Nowgong, Golaghat, Sibasagar, Dibrugarh and Saikhowa Ghat. The other road runs through the North bank, Guwahati to Lakhimpur through Tezpur. In spite of the recommendation of the civil and military authorities, Assam Trunk roads received less attention as it deserved until the Jayantia rebellion (1862). High Rose, the Commander-in-Chief of army recommended the early construction of military roads both across the hills along the valleys of Brahmaputra and Surma.

In February 1864, Government of Bengal sanctioned two main lines of communication in North-East India, one highway run along the entire length of the valley Dhubri (Assam) in the Western frontier of Assam to Sadia in the East and the other road to connect Chattak on the Surma River with Guwahati passing through the highland of Shillong and Yado (Barpujari, 2007: 294). Apart from the trunk roads the most important feeder roads were, the road from Tura in the Garo Hills to the bank of Brahmaputra, the road from Guwahati to Disangmukh and the Dhodoar Ali in the Sibasagar District (Allen,1905: 90).The progress of work however, slow on account of both financial constrain and acute shortage of local labours. Since, there was hardly any landless peasant, agricultural wage-labourers were also very few in the province. It was only during the slack agricultural season that some local rural labourers were available for hire. Therefore, Public Works Department had to depend primarily on the imported labourers and contract expired tea garden labourers. On the other hand, the undulating topography, frequent flood, heavy rainfall, alluvial soil requiring stone for metalling which had to be carried over long distance which slowed down task of the road construction work (Kumar, 2008: 211).

By 1903, 7,799 miles of roads had been constructed in Assam but of this, two per cent were metalled road (Goswami, 2012: 170). In the districts of Brahmaputra Valley heavy rainfall necessitated embankments for the roads and the lack of road metal in the region combined with heavy cost in transporting construction material from the other regions made it impossible for the most of the road to be metalled. Therefore, all practical purposes, the majority of the roads in Assam could be used only for a few months in a year. In the rainy season, it was almost impossible to use the non-metalled roads in most of the areas (Antrobus, 1957: 72). In 1911-12, the total length of road in the whole province was 3,559 miles including 375 miles of metalled roads.

From 1860 s onward the British Government however, started building some roads and opened lines of communication for the movements of troops throughout the whole year, for conducting military expedition into some tribal areas. It was for this reason that the development of infrastructure consisting of roads-ways, river-ways and the railways was undertaken on a large scale to meet the needs of the tea industry in the mid-nineteenth century.

In 1868, Public Works Department was established for the purpose of carrying out the programme of public work. In 1880, Local boards were constituted to build roads for meeting the local needs. For such works resources of the local board came from their share of "half the proceeds of a local rate of one-sixteenth the annual value of all landed property", tolls collected from the pounds and ferries and some government grants (Kumar, 2008: 211).

Over the years the local boards became actively involved in development works at the local-level in the fields of road development, education, public health, medical service etc. Towards the close of the year 1920-1921, there were ten Local Boards in the whole of Assam which having elected a non-official chairman and later three more Boards followed suit (*ibid.*). In 1853, James Mill reported that there was only one road in the district of Cachar which had not been repaired for many years. In 1872, according to the list of the roads returned by the Deputy Commissioner there were only two roads in the district of Cachar under the supervision and administration of Public Works Department. On the other hand, the total number of roads under the supervision of local administration counted to be 15 in the district. Assam Administration Report of 1875-1876 thus returns the mileage of the means of communication in Cachar district which includes navigable rivers, 361 miles, Second-class road 208 miles and Third class road 300 miles (Hunter, 1879: 231, 232).

Even up to the end of the nineteenth century, complaints about the roads were a regular subject in all the reports of the Indian Tea Association. In 1892, the chairman of the Indian Tea Association complained that the district of Cachar could not boast of a single road of a mile in length. In fact, in the whole of the Assam, the proportion of metalled road to non-metalled roads was small. In the Surma Valley, communication by water completed with communication by road during the rainy season and that was why, less outlay was incurred on the roads than it was otherwise necessary (*ibid.*).

In 1911, the Indian Tea Association urged the British Government of Assam to give importance and to the development of Road ways. Government realizing the value of the recommendation of the Indian Tea Association in 1912, a sum of Rs. 24,00,000 were distributed as special grant for the development

Table 11.1: Roads in Cachar District under Local management returned in 1873

Serial No.	*Name of the Roads*	*Area Connected by the Road*	*Length of the Road (miles)*
1.	Barkhola Road	Tarapur, Jainnagar, Barkhola with Silchar	9
2.	Durganagar Road	Rangpur, Bahadurpur, and Udarband with Silchar	7.5
3.	Chatla Hamor Road	Ambikapur, Maharpur, and the Adjoining tea Gardens with Silchar	18.5
4.	Nagdirgram Road	Sonapur and Nagdirgram	12.5
5.	Manierkhal Road	Tea gardens of Manierkhal and Nagdirgram	9
6.	Tarapur Road	Tea gardens of Tarapur, Lahak with Banakshandi	7.5
7.	Silchar Municipal Road	Area of Silchar Town	8
8.	Baladham Road	Kamaranga and adjoining tea gardens with Lakhimpur	10
9.	Sonai Road	Tea gardens of Hatikuri and Binakandi	11
10.	East Hailakandi Road	Somaikona and Matiguri with Hailarandi Subdivision	12
11.	Katakhal Road	Golachira tea gardens and Sudarsanpur	8
12.	Mona Road	Gongpur, Dumkar with Rajeswarpur	8
13.	Hailakandi Road	Panchanagar, Rafinagar and Kashinagar	13
14.	Masimpur Road	Barkhola and other tea gardens with Masimpur	10
15.	Natwanpur Road	Natwanpur and tea gardens of Jainagar	22

Source: Hunter, 1879: 231, 232.

of road transport in Assam. It was only in 1922 first passenger bus service started between Jhanji and Jorhat in the Upper Assam and since then regular passenger service has been in vogue on this important road. Thus, prior to the initiative of the start of motor service in 1922 passengers and mail to Shilliong were carried on by *tongas* and horses. It was during the Second World War that the road construction was accorded top priority to meet the defence need. Towards the end of 1942, when there was acute shortage of transport capacity owing to requisitioning of vehicles for military operation, Assam Transport Organization, a private enterprise was set-up with a few and 1,100 employees within six months. But after the end of the war this organisation declined and therefore, Assam Government set-up a committee to go into its affairs. When this Committee recommended the retention of Assam Transport Organisation, Government decided to remodell it and named it as Assam State Transport with effect from 1948 (Medhi, 1978: 42, 43).

British Commercial Interests

The development of tea industry and the other modern industries in Assam compelled the Government for a developed transport system. The Industrial revolution of England further pressurised Government to find market to sell their industrial finished products. They also wanted to acquire an uninterrupted supply of raw industrial materials for their factories and their best choice was their colonies (Stephenson, 1916: 3). Therefore, British Government in India took interest to connect the hinterlands with the roads, rail and water ways for the easy transport of the rawmaterials to balance the pressure created by the Lancashire and Manchester textile mills (Thambi, 2005: 48).

The initial advocates for developing railways and road ways were the mercantile firms in London and Manchester with trading concerns in India. To meet their industrial need it was important to link the hinterlands with the ports for batter transportation. The importance of the development of road and the other means of transport from the Economic standpoint of view came to be realized from the middle of the 19th century. The main objectives were the exploitation of the natural resources of India in general and Assam in particular (Chandra, 1979: 14). Government constructed roads just to carry the forest wealth in the hilly areas and these for a temporary purpose therefore repairing was done not regularly. Therefore, the very purpose of the development of transport system in Assam during Colonial period was to serve the British Commercial interests. The Indian Nationalists also criticized the colonial pattern of finance said that the railways and the roadways had not been coordinated with India's industrial needs (Chandra, 1988:95).

British rule in Assam brought certain major changes both in the field of economy and administration. With the development of tea, oil and coal industries in Assam, British Government adopted a forward policy in order to bring all the neighbouring hill territories under their political domination. But after the consolidation of British control over the Naga Hills, British abandoned their plan to develop the area and introduced a series of rules and regulations which isolated the entire Naga Hills. The Inner Line Regulation virtually ended the commercial and kinship relation between the people of the plain and the hill. As the British Government consolidated their control over the markets of Assam, trade through the *duar* or doors gradually petered off and the primary outlet that existed was the western doors or the *chowkey* at Hadira in the bank of the Brahmaputra. In 1833, Hadira Chowkey was brought under the British control and the free movement of the goods was introduced there by ending custom duty. Government constructed trunk road from Sadia to Dubri, introduced rail and river navigation in Brahmaputra. After the end of Hadira Chowkey Assam Bengal trade took a new turn and after the growth of tea industry, tea became a principle export commodity in the province. Thus, the introduction of road, rail and river traffic along with new mercantile interest, Assam's trade relation with the bordering states radically changed. Assam trade with Bhutan and Nepal declined considerably and relation with Bengal took a new dimension.

British Strategy of Road Development in Assam

During the wake of the development of tea industry in Assam some roads had been built by the British Government, but their intension was not to develop the economic status of Assam, rather they intended of their own mercantile gain. The development of Tea Garden, Road transportation and even education system was developed in Assam in order to flourish Christianity in the remotest corner of Assam (Chandra, 2005: 58). The initial thrust of the Government, after the annexation of Assam, was of maximising land revenue by increasing its manifold. Therefore, to set the economy on the path of capitalistic development, the British colonialists sought to link Assam's village markets with World market to maximise their benefit. A major bottleneck in exporting the agricultural products from the province was the absence of a reliable means of transporting bulk goods.

The basic criterion for choosing road alignment during British colonial rule was military feasibility, for providing speedy troop movement throughout the whole year, in case of incursion of the frontier tribes. In 1940, alarming reports of Burmese attempts to open a road through the Naga Hills made it essential to improve the line of communication between the Motok Country and the troops reserve at Sibasagar. Steps were taken to repair the *Bar-Baruah Ali*, the high road run from Sibasagar to Dibrughat through *Chauki-Hat* on the *Buri Dihing*. This road had a grand political importance, as affording the means of moving troops and of controlling the Naga peoples (Hilaly, 2007: 151, 152). Repairs were also undertaken on the road from Dibrugarh to Saikhowa to facilitate the easy movement of the Assam Light Infantry, which has to guard the extensive frontier. To complete the line of communication along the frontier posts, a road was constructed between Dum-Duma and Makum also. A small stretch of road between Golaghat and Nagorah was constructed, as because it was much frequently raided by the Angami Nagas.

In the wake of the development of tea industry, good number of roads had been built to link the tea gardens to landing places on the Brahmaputra, as well as the steamer, wherefrom country boats to Calcutta ship their products. Government also emphasised the need for road building in the hills areas, to establish markets and develop them, "as roads and markets ought very speedily to create a social revolution in the hills". A road was opened from Goalpara to the Eastern hats of *Jirah, Damrah* and *Nibari*. It was of great importance to the merchants of Goalpara as there was a large trade in cotton (*ibid.*).

Conclusion

In the context of Colonial Assam, road construction began as a part of war preparation for the First Anglo-Burmese war. Every treaty signed between the East India Company and the local rulers after the Burmese war typically implied the need for road construction such as treaties with Cachar, Jayantia, Khasi Hills and Manipur. The military department built the roads, maintenance was transferred to the civil department, who in turn dependent on the local Rajas to get the job done. The Government used the non-local labour and local *Cooli* labourers for the construction of the roads and the local *Rajas* acted as the agent for the supply of the labourers. The growth of the tea industry necessarily led to the growth of the transportation and infrastructural development in Assam. The industry which was totally dependent on an external market required the basic routes to and from Calcutta. Therefore, the roads which were built in Assam in the wake of the development of the tea industry, was basically to facilitate, the movement of the bulk goods and the labour imported from the other states. Hence, most of the roads in the tea growing districts developed considerably and all these connected the tea gardens and the river ports, to transport goods to and from Calcutta.

This is further proved by the fact that it was only the tea growing districts of Assam where hinterlands were connected by the roads. But in the districts of Goalpara and Darrang where there was less potentiality of the production of tea, the road condition was very worse. There were neither metalled roads nor the non-metalled roads were constructed for connecting the hinterlands. But it was reverse in case of the district of Cachar and Lakhimpur where good number of metalled and non-metalled roads had been constructed only because of considering the potentiality of the growing tea industry in these districts. The road development policies of the British Colonial Government in Assam was formulated to serve the trade and mercantile interests of the Government not to provide better transport facility for the welfare of the common people.

References

Antrobus, H.A. (1957), *A History of Assam Company,* Edinburg.

Barpujari, H.K. (Ed.), 2007, *The Comprehensive History of Assam,* Guwahati: Publication Board of Assam.

B.C. Allen (1905), *Gazetteers of Bengal and North-East India,* Calcutta: 224.

Bhushan Chandra (2005), *Assam, Its Heritage and Culture,* Delhi: Kalpaz Publisher.

Bhuyan, S.K. (1949), reprint 1974, *Anglo-Assamese Relations, 1871-1926,* Guwahati.

Chandra, Bipan, (1979) *Nationalism and Colonialism in Modern India,* Delhi: Orient Longman Publications.

Chandra, Bipan, (1988), *India's Struggle for Independence,* New Delhi: Penguin Books.

Gait Edward, (2008), *A History of Assam,* Fourth Reprint, Delhi: Surjeet Publications.

Goswami, Priyam, (1999) *Assam in the Nineteenth Century: Industrialization and Colonial Penetration,* Guwahati: Spectrum Publications.

Goswami, Priyam, (2012) *History of Assam, from Yandaboo to Partition,* Guwahati: Orient Black Swan Publication.

Hillaly Shah, (2007) *The Railways in Assam 1885-1947,* New Delhi: Pilgrims Publishing.

Hunter, W. W., (1879) *A Statistical Account of Assam,* Delhi.

Kar, M., (1997) *Muslims in Assam Politics,* New Delhi: Vikas Publishing House.

Kumar, A. Roy, (2008) *Society Politics and Development in North-East India,* New Delhi: Concept Publishing Company (P.) Ltd.

Lahiri, Nayanjoti, (1991), *Pre-Ahom Assam,* Delhi.

Medhi, B.S. (1978) *Transport System and Economic Development in Assam,* Guwahati: Publication Board of Assam.

Stephenson James (1916) *Principrls and Practice of Commerce,* London: Pitman & Sons Ltd.

Thambi, Madhabi (2005) *India and China in the Colonial World,* New Delhi: Esha Beteille Social Science Press.

Section C
Region and Beyond

12

Role of Public Works Department in Making Road in Assam

Tarun Dutta

Introduction

The present road system has its origin in remote past as in case of all old settled lands where the ancient trails serving as avenues of intercourses, but with fuller settlements, the trails were improved, extended, abandoned according to the needs of the settlers. Roads originated for economic reasons, related to the need to move food and other goods from one point to another. Early transportation focused on moving food from a hunt or a harvest to the places where people lived. Trails evolved from prehistoric animal paths, and early humans carried or dragged their loads along these paths. As humans learnt to domesticate animals, they transferred their loads to pack animals, such as horses, mules, camels, llamas, elephants and dogs.

During the period from 1900 to 300 BC, four trade routes, known as amber roads, ran across central and eastern Europe. Amber was an important ingredient in primitive and medieval medicine and was also used for statuary and jewellery. One of the first pre-planned roads was the Persian Royal Road, built by Darius I in 500 BC in what is now Iran. The Royal Road was about 2,400 km (about 1,500 mi) long and stretched throughout ancient Persia. The road was constructed for royal use, and it allowed Darius to keep informed, to convey orders, and to transport goods needed by the royal court. The road followed the shortest and most efficient route. It bypassed some of the largest towns—in part, for the sake of following the shortest route, but also because large towns had the potential to rebel. An uprising in a town near the road could have threatened the security of the route or prevented troops from being dispatched.

The Chinese Silk Road, established around 100 BC, was a series of land routes connecting ancient Rome and China. Caravans of traders travelled the Silk Road, exchanging goods and ideas. The road extended about 6,000 km (about 4,000 mi).

The ancient Romans built an extensive and durable system of roads to serve their needs. The Roman road-building era began in 312 BC. The roads provided economic and military access from Rome to distant parts of its far-flung empire. The first road constructed was the Appian Way, which led from Rome to Brundisium (now Brindisi), a port in what is now southern Italy. The Appian Way was the main route to Greece, and it ran over 560 km (350 mi). A second road, from Rome to Naples, provided the first stage of the route used by troops headed to Africa. Roman advances in road-building techniques included preparation of foundation soils and base courses, brick paving, and most importantly, provision for adequate drainage.

In the western hemisphere, road building evidently reached a high state of development in the 15th and 16th centuries under the Aztecs in Mexico, the Maya in central America and the Incas in south America. The Spanish conquistadors, in turn, constructed a series of *Caminos Reales* (royal roads) during the 16th and 17th centuries. These were the first American highways, and they stretched from Veracruz, Mexico; northward to San Francisco, California; and Santa Fe, New Mexico; and eastward to Saint Augustine, Florida.

Important advances in road-building technology came in the eighteenth century. Pierre Trésaguet in France and Thomas Telford and John Loudon McAdam in Scotland introduced improved methods for creating roads made of stone. McAdam's method stressed the need to keep the subsoil dry by including adequate drainage and waterproof covering. The term 'macadam' originally referred to McAdam's road-surface design, in which crushed or broken stone was mechanically locked together by rolling a large weight over the surface. MacAdam was later used to describe other types of bound surfacing, including water-bound MacAdam, in which the crushed stone is cemented together with an application of water. In penetration MacAdam, bituminous material sprayed on the compacted rock binds the rock together.

The passage of Indian roads is as old as its civilisation. When the Harappa civilisation was existed in the bank of Indus and its neighbouring areas, a well-furnished road connected the areas. In viewing good communication, fast transportation and better defence, early rulers of India maintained a good communication roadway within their territory. They had even maintained good communication with neighbouring friend countries. The great Mughals, who ruled almost the entire Indian sub-continent, built a good number of roads. The objective was to obtain information fast about running of the administration and security arrangements.

East India Company, the mercantile company of England, after landing in India, realised the need of good transportation system for their imperial trade and for that purpose they built a good number of roads. In 1835, the first railway line was established, which connected Bombay (Mumbai) and Thane, and later it was expanded to most of the region to obtain raw materials and deliver goods manufactured in Europe in the regions.

Public Works Department and Road Making in Assam

With the establishment of the permanent PWD in 1868, a number of district and sub-divisional headquarters were created. The headquarters and sub-divisional offices were communicated by construction of new roads. The first notable work done by the Government was the construction of Assam Trunk Road. By 1871, the Assam Trunk Road was completed from Guwahati to Agia in Goalpara district. Since, the introduction of Chief Commissionership in Assam, the Trunk Road had been extended to join the Bengal Road system.[1] From Goalpara to Singimari, a distance of 102.4 km, and another road of 41.6 km between

Dhuburi and Kerbari, were also open to connect Cooch Behar road. However, the roads were used only by cattle dealers, elephant catchers, labourers of tea garden and pilgrims. Only the bullock carts were used in the Assam Trunk Road for short distances. The other roads were not passable throughout the year. The roads in many areas passed through the cultivated tracts of land and jungles. All goods traffic of the Brahmaputra Valley region was usually carried by the river. During these days, North Trunk Road was less important so far as flow of traffic was concerned between Assam and Bengal, because the South Trunk Road had direct connection with the road system of Bengal. The north bank of Dhuburi town developed into an important trade centre in which the traders of Assam, Eastern (Bangladesh) and Northern Bengal could meet and interchange their products.

Growth of tea industry played a crucial role in the development of roadways. The tea garden also assisted in road communication and development. As early as 1845, the Assam Company claimed, with some exaggeration, that they had opened or repaired some 1,280 kilometres of public road, had erected 266 bridges and established several ferries across the rivers.[2] The industry, for first two to three decades, solely depended on the development of road on its own effort. The roads were mostly linked the gardens to landing places on the Brahmaputra, as well as to steamer wherefrom the country boat shipped the product. The roads linking the gardens and the villages materially improve the condition of imported labourers and it affords the cultivators, increased facilities for disposing of the produce in their lands.[3] To link with landing places (*Ghat*) on the Brahmaputra and the tea gardens, a portion of the *Dhai Ali* between Sibsagar to Buri Dihingmookh; *Barpatra Gohai Ali* from Sibsagar to Disang, and the *Sologuri Ali* from Disang to Buri Dihing were constructed.[4] Emphasis was given to build road with hills to establish market and trade development with the hill people, as viewing that roads and markets ought very speedily to create a social revolution in the hills.[5] Between Udalguri and Mongaldoi, a road was constructed with an object to the trade with Bhutias so that they can come down during the Annual fairs.[6] With the objective of the cotton trade of Goalpara, a road was opened from Goalpara to the eastern hats of Jirah, Damrah and Nalbari.[7] In Assam, a large number of roads were lying unused in that time. After the stabilisation of political condition some main roads were repaired for commerce and public use as early as in 1835. The principal roads of *Kamroop* were repaired, the great western road from Guwahati to Goalpara was also repaired and *Gohai Kamal Ali,* between Manas and Buri Nadi, was restored. The high bund road extending from Guwahati through East Dimarua to Raha and from Nagaon to Jorhat and up to Sibsagar, was restored.[8] However, most of the roads were damaged shortly due to its non-permanent nature of work.

In between 1870 and 1880, many roads under PWD and Local Boards were repaired and reconstructed and made fit for the wheel traffic. Some of those roads in the Brahmaputra Valley in 1873[9] are shown in Table 12.1.

Table 12.1: Roads in the Brahmaputra Valley in 1873

Districts		*Roads*	*Length in km*
Goalpara	1.	Assam Trunk Road—from Agia to Guwahati	130
	2.	Goalpara to Singimari	102
	3.	Dhubri to Kherbari	42
Kamrup	1.	Guwahati to Kalangpar (A part of Assam Trunk Road)	24
	2.	Guwahati – Shillong	104
	3.	Guwahati – Haflong	131
	4.	Amingaon – Hajo – Nalbari	51
	5.	Kamalpur – Jhargaon	32
	6.	Rangia – Nalbari – Barama	38
	7.	Barama – Tamulpur	32
	8.	Bajali – Bhabanipur – Barpeta Road	38
	9.	Barpeta – Baniaghat	32
	10.	Barpeta – Roha	16
	11.	Palashbari – Gohainkant	13
	12.	Gohaikant – Khanajan	16
	13.	Kayirapur – Bardwan	22
	14.	Amchang Road	8

(Contd.)

Table 12.1 (*Contd.*): Roads in the Brahmaputra Valley in 1873

Districts		*Roads*	*Length in km*
Darrang	1.	Guwahati (North Bank of the Brahmaputra) to eastern boundary of the Darrang district	253
	2.	Orang – Odalguri	24
	3.	Odalguri – Bengbari	14
	4.	Odalguri – Bhairabkunda – Amratal	18
	5.	Bengbari – Mangaldoi	40
Nowgong	1.	Nowgong – Guwahati	91
	2.	Nowgong – Dhing river	86
	3.	Kaliabor – Silghat	6
	4.	Nowgong – Doboka	38
	5.	Nowgong – Laskhoyaghat	27
	6.	Kahargaon – Kaliabor	56
	7.	Puranigudam – Bamunji	10
Dibrugarh	1.	Rongagarh Road	40
	2.	Rongdoi Ghat Road	19
	3.	Sologuri Road	32
	4.	Hugrijan Road	13
	5.	Rajgarh Road	5
	6.	Monuata Road	11
	7.	Doom Dooma Road	32
	8.	Moria Ali	8
	9.	Bura Gosai Road	24
	10.	Mohamaria Road	24
	11.	Rangmalla Road	8
	12.	Bayaloni Road	8
	13.	Tippum Road	8
	14.	Nadwa Road	11
	15.	Dhodar Ali	32
	16.	Makum Road	32
Lakhimpur	1.	Garmur Road	24
	2.	Kachujan Road	58
	3.	Pathalipam Road	29
	4.	Dhakuakhana Road	16
	5.	Kherkatiya Road	58

Local Bodies and Progress of Road Making in Assam

With the establishment of the Local Boards in Assam in 1880, the real progress of the roads of local importance was made.

It was felt very difficult to maintain trade between the Brahmaputra and the neighbouring hills. Therefore, the government rightly emphasised the need for road building in hill to establish markets and develop trades. In an alluvial plain, the valley is crisscrossed by innumerable streams and streamlets. This has necessitated commutations and maintenance of countless bridges and culverts at close intervals and at a very heavy cost. Consequently, the road running from Harigaon to Mankachar in the border of the Garo hills *via* Goalpara, was constructed in 1875. The Shillong-Guwahati road was made fit for wheeled traffic in 1877. The other hill roads were Golaghat-Nagahills (107 km); Nagoan-Nagahills *via* Manipur Road and North Cacher (135 km) and Guwahati-Bholaganj in Sylhet *via* Khasi and Jaintia hills (208 km). In reality, these were mere footpaths.

From 1896 onwards, cart traffic on the roads of the Brahmaputra Valley played an important role in the trades of Assam Valley with Manipur and neighbouring areas. Since the beginning of the twentieth century, the valley had a considerable mileage of roads under the PWD and the Local Boards. The expenditure by the Local Boards and by PWD on road communication (both original and repairing work) rose from Rs. 90,416.00 to Rs. 8,30,723.00[10] and from Rs. 46,549.00 to Rs. 474,465.00 respectively in 1924-25.[11]

In order to exploit its resources, Assam had made great efforts in the direction of road improvements in the recent years. A comprehensive programme chalked out in 1928 to construction and improvement of the roads, brought not only a better co-ordination between rail, river and roadways, but also a good network of roads in the valley region. Occupation of Assam province by the colonial power made changes in the structure of the communication. They developed the means of communication in viewing military and commercial feasibility. Their objectives were speedy movement of troops, incursions of frontier tribes as well as to resume the case of Burmese attack. They were emphasising to restore portion of great roads of Upper Assam[12] and undertook the repair of *Bar Baruah Ali*, the high road from Sibsagar towards Dibrugarh through Chokey Hat on the Burhi Dehing, a portion of high roads between Jaipore and Sibsagar. The road had great political importance as affording the means of moving troops and of controlling the Nagas, as well as highly valuable as one of the roads of internal communication on a line, by which much traffic in salt is conveyed.[13] From Dibrugarh to Saikhowa and Jaipore portion of road was repaired to facilitate movement of the Assam Light Infantry, which had to guard the extensive frontier.[14] The Bund road from Sibsagar to Dikhu Mookh was of highest utility for military and political point of view.[15] To conduct a Naga operation, Captain Butler, the principal Assistant of Nowgong, urged to construct two roads, one to be from Nowgong *via* Samuguting to Dimapur and the other from Golaghat to Dimapur.[16] To complete the line of communication along the frontier post, a road was constructed between Doom Dooma and Makoom.[17]

To colonise in the neighbouring area, British made efforts to build road to connect those areas. In 1826, efforts were made to establish an external communication with Sylhet through the territories of Khasi Chief, Khyrim and Mylliem. The work commenced in 1829, but due to disturbance and murders of the in charge, Lietenant Burlton and Beddingfield, the construction stalled. The project was restarted in 1832, after the annexation of Khasi Hills, as a military feasibility route. To connect all the districts of Assam, the 'Trunk Road' project was formulated in 1866 with the design of making it a first class metallic and embanked highway, twenty feet wide and aligned in a manner to make it possible to convert it into a railway.[18] After the accessing and taking stock of the available funds, the idea of a railway formation for the Trunk Road was given up and orders were issued to continue work on a smaller scale of a sixteen feet top width.[19] The project was stalled due to the paucity of funds. Fragmentary portion of the old roads throughout the province were constructed to form a continuous trunk road, and finally extended towards Bengal.[20]

Again, the extensive works on roads communication were programmed in 1935 and 1937-38[21], but a large number of project remained to be completed owing to the outbreak of the Second World War.

Second World War and Road Making in Assam

With the outbreak of the Second World War in September 1939, the transport difficulties became more acute and the region needed a secured road communication. Consequently, the construction of roads was given more priority as per the needs of defence. Accordingly, the Goalpara-Saikhuwa route and Makum-Lekhapani road were given first priority for development for smooth running of military traffic.

The construction and repair of these roads were basically for military purposes and most of them unmetalled. These were used only for passage of military troops and elephants and its nature was purely temporary. As the political situation gradually stabilised, military road construction was no more required and it gradually moved to commercial purposes. In the time of Second World War, the roadways in Assam were only for trade interests and only a few roads were for defence purposes. Trade and commercial linkage with the neighbouring countries were tried to establish by building rail and roadway, but all efforts did not materialise due to various constraints. When the strategic importance was felt and the necessity of supply road emerged due to Japanese blockade of China, the linking road building work from North-east India was initiated.

Notes

1. W.W. Hunter, 1879, *A Statistical Account of Assam*, Vol. II, Trubners & Co., London, pp. 72-74.
2. A. Guha, *The Geography Behind the History*, p. 160.
3. Report of the Commissioners Appointed to Enquire into the State and Prospects of Tea Cultivation in Assam, Cachar and Sylhet, 1868, p. 101.
4. A.J.M. Mills, *op. cit.*, Appendix F-1-23, List of principal roads in Sibsagar District.
5. Bengal Government papers, nos. 1-3, file no. 15/18(1873).
6. Letters Issued to Government, Vol. 16, no. 320, 10 August, 1852.
7. Goalpara papers 1873, file no. 981, serial no. 122, 28 February, 1873.
8. Bengal Judicial Consultations (Criminal), no. 4, 13 October, 1835.
9. W.W. Hunter, *A Statistical Account of Assam*, Vol. I, pp. 72-74 and vol. II, pp. 56-378.

10. G.O.A., Resolution on the working of Local Boards in Assam (1924-25).
11. G.O.A. Report on Public Works Department (1924-25 and 1925-26).
12. Letters Issued to Government, Vol. 9. January-September 1840, no. 23, 28 January, 1840.
13. Letters Issued to Government, Vol. 9. January-September 1840, no. 147, 28 August, 1840.
14. Letters Issued to Government, Vol. 10, no. 131, 9 March, 1852.
15. Bengal Judicial Consultations (Criminal), no. 40, 19 September, 1842.
16. Letters Issued to Government, Vol. 15, no. 112, 21 September, 1850.
17. Letters Issued to Government, Vol. 15, no. 152, 28 November, 1850.
18. W.W. Hunter, 1975, *A Statistical Account of Assam*, vol. I, rpt, Delhi, p. 55.
19. *Ibid.*, pp. 47-48.
20. S. Hillaly, 2007, *The Railways of Assam, 1885–1947*, Varanasi, p. 59.
21. G.O.A. Report on the Administration of Assam Public Works Department (1939-40), pp. 7-14.

13

Development of Road Transportation in North-East India

A Case Study in Dibrugarh and Tinsukia Districts Region, Assam

Sangeeta Boruah Saikia and *Narendra Kr. Das*

Introduction

Road transportation has been a major bottleneck for the economic development of a country. Assam may be called the gateway to the other six states of North-East India and as such the improvement of road transportation facilities should be the first consideration in all planning and development programmes of Assam.

The present layout of the road transport network of the Dibrugarh and Tinsukia Districts more or less has developed over the framework laid down by the colonial rulers during their regime. Ahom rulers contributed a lot towards the development of road transport but except for a few of them these roads became useless to the British. The tea industry gave incentive for rapid development of road transport system of the Region. Construction and development of roadways linking the tea gardens with the trunk road has been started since 1868. The creation of the Public Works Department in 1868 and the local Boards in Assam in 1880 was a significant step taken by the British to expand and develop the road network of Assam as well as Dibrugarh and Tinsukia Districts.

The study will attempt to find out the development scenario of road transportation in Dibrugarh and Tinsukia Districts which is very much important for regional development of a region.

Study Area

Dibrugarh and Tinsukia Districts, previously constituting the Dibrugarh District is an emerging urbanized area next to Kamrup

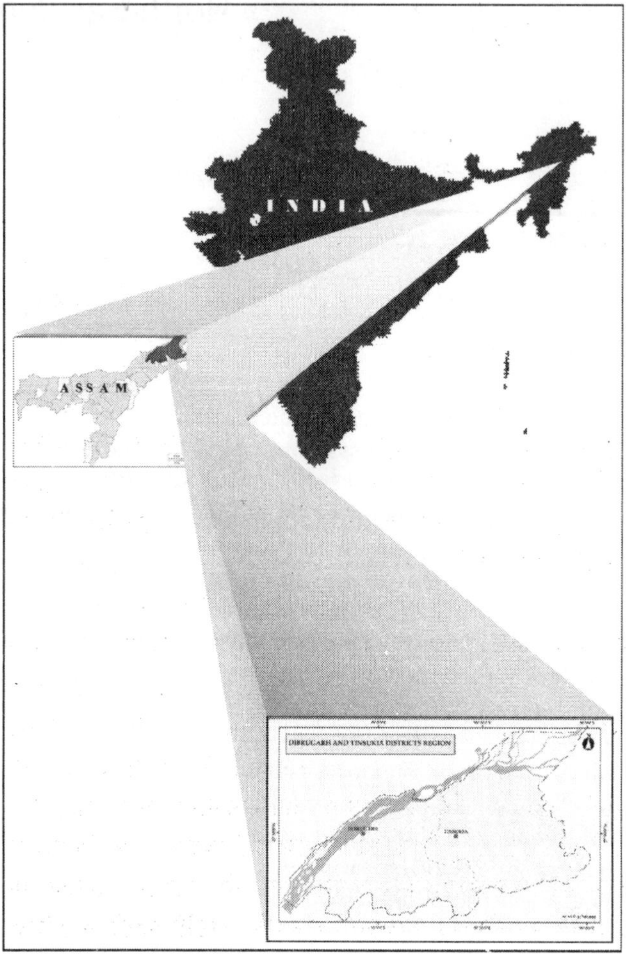

Source: www.mapsofindia.com

Fig. 13.1: Location of Dibrugarh and Tinsukia Districts Region

District. The region is located in between 27°5′ N and 27°58′ N latitude and 94°35′ E and 96°0′ E longitude in eastern most part of Assam south of the Brahmaputra. It is the area where modern life started first in Assam through the establishment of tea gardens and coal and oil industries. First railway lines and trunk roads were initiated in this area during colonial era laying ground for modern urbanization. It covers a total area of 7171 km^2 consisted mostly of plains with scattered hill and hilly area in its eastern and southern margins bordering Arunachal Hills.

Aim and Objectives

(i) To find out the past scenario of road transportation in Dibrugarh and Tinsukia Districts.

(ii) To see the present status of road transportation in this Region.

(iii) To understand the problem and future prospects of road transportation in this Region.

Data and Methodology

The data for this works are collected from census reports and *Gazetteer of India*. The topic for investigation is taken through intensive library work to find out relevant informations in books, journals, articles, monographs etc. and also assembling personal experiences, theoretical knowledge and imaginations, opinion, observation in the field. GIS softwares are used for preparation of maps for comparison.

Results and Discussions

Old Time Road Transportation

Information of the old time overland trade routes of the undivided Lakhimpur district is very mearge. Several old evidences shows that as early as in 200 B.C. there were land routes connecting China, through Burma with Kamarupa. Some of these routes stretched through this region. It is evident that the undivided district of Lakhimpur had been a doorway to the plains of Assam in early times when the tribes of the Mongoloid stock migrated through the steep hills of Arunachal Pradesh. Traveller who traversed the land during the 18th and 19th centuries left some interesting accounts of the trade routes of this region." McC'osh refers to as many as five roads leading from Sadiya, the frontier stations of the Brahmaputra valley into Tibet, or China proper. They are: the pass of Dibong, the Mishmi route, the Phungan pass to Mansee and China. Through this route, Shan invaders came to the Brahmaputra Valley. In 1816 during the Burmese invasion, some 6000 Burmese troops and 8000 auxiliaries crossed the Patkai into Assam. During the reign of Koch king Naranarayana of the present Lakhimpur district was constructed. The king's brother Gohain Kamala supervised the work and the road was known as the "Gohain Kamala Ali".

Ahom rulers contributed a lot towards the development of road transport but except for a few of them these roads became useless to the British who were only concerned with the extraction of natural resources from this territory consequently, with the annexation of Assam to the British rule through the Treaty of Yandaboo in 1826, the erstwhile road transport structure of the state began to change slowly. The creation of the Public Works Department in 1868 and the Local Boards in Assam in 1880 was a significant step taken by the British to expand and develop the road network of Assam with a view to collecting resources of the state and then to dispatching the same to their own land through Calcutta.

After the advent of the British, there was marked improvement in transport and communications in the undivided district of Dibrugarh. British expanded and developed the road network in Dibrugarh and Tinsukia Districts Region with a view to collecting resources of the state and then to dispatching the same to their own land through Calcutta. In 1902-03, there were 23 miles of metalled and 723 miles of unmetalled roads, maintained either by the Public Works Department or the Local Boards, the last named authorities being responsible for about five-sevenths of the total length (*Gazetteer of Lakhimpur*, 1905). The real development in respect of the construction of new roads did not take place until the establishment of the Road Board in 1926-27.

After independence, the Government of Independent India has been emphasizing more on the development of road transport facilities within the state in order to expand the British laid network so that the passengers and goods can be transported even from the interior areas. A significant proportion of investment allotted for the development of road transport system during the First and the Second Five Year Plans (Bhagabati, Bora and Kar, 2001).

At the end of 1961, the total length of roads under the management of the Public Works Department was 1268.177 kms. of which 215.34 kms. was blacktopped, 865.316 kms. gravelled and 270.179 kms. earthen. There had been considerable extension of the state roads during the last few years. The Department had taken-up improvement and expansion of the existing single lane carriage way to two lanes and reconstruction of weak and narrow bridges and culverts. Some of the old roads have been extended and a few others had been newly constructed.

Formerly, the village roads were maintained by the respective Local Boards. But with the introduction of the Assam Panchayat Act, 1959, the Local Boards were abolished and since then rural communication had been taken over by the Community Development Blocks. Almost all the village roads are unmetalled and became motorable only in the dry season. Proper maintenance of village roads in the flood affected areas is a colossal task. Formerly, these roads were repaired by the local people. The Tea Estates also have many roads constructed primarily to transport the garden products. Broadly speaking the condition of these roads is much better than that of village roads.

The position as in 1970-71 is shown in Table 13.1.

Table 13.1: Dibrugarh and Tinsukia Districts Region: Length of Roads under the P.W.D. in Dibrugarh and Tinsukia Districts, 1970-71

(in km.)

Name of Districts	*Black topped*	*Gravelled*	*Earth*	*Total*
Dibrugargh and Tinsukia	402.20	36.35	44.33	1382.88

Source: Statistical Handbook of Assam, 1971, p. 107.

Present Status of Road Transportation

Dibrugarh and Tinsukia Districts Region is connected with the rest of the state by all the four modes of communication, i.e. road, railway, river and air. The National Highway Nos. 37 and 38 form the artery of road communication in the region. A number of black topped and earthen gravelled roads branch off from these two roads and form network of road system. The National Highway No. 37 which was formerly known as the Assam South Trunk Road, enters the region from Sibsagar and runs from Moran in north-eastern direction up to eastward to Saikhowaghat which stands on the bank of the River Luhit. The Makum Lekhapani Road now known as the National Highway No. 38 takes off from Makum and reaches Ledo through Digboi and Margherita.

The State transport bus service and private bus services connect this region with Sibsagar, Jorhat, Golaghat, Nagaon, Guwahati to the west, Chabua, Tinsukia, Digboi, Ledo, Duliajan, Naharkatia, Namrup, Sonari etc. to the east and Saikhowa, Chapakhowa, Roing, Tezu, Changlang, Khonsa to the north-east. Beside these, taxi and trucks are playing a major role in transporting men and goods to and from in the region. The region has 0.5 km per km^2 road density with total length of roads 3503 km (Table 13.2).

Table 13.2: Dibrugarh and Tinsukia Districts Region: Length of Roads 2015

(in km.)

Type of Road	*Length*	*National Highways*			*P.W.D. Road*		
		Under Construction	*Surfaced*	*Total*	*Black Topped*	*Earthen Gravel*	*Total*
1. State Highway	218						
2. Major District Road	334	100.257	309.599	409.856	2146	1357	3503
3. Rural Road	2836						
4. Urban Road	115						
Total	**3503**						

Source: Transport Research Wing, Ministry of Road Transport and Highways.

There is no established pattern of form of road network in Tinsukia town except few major roads most coming-up in an organic way even without least consideration of hierarchy. In addition to this improper geometry, poor maintenance has made the level of service very poor.

City roads in Dibrugarh are smooth and well built-up with street lights. However, keeping in view the rapidly increasing number of commercial and private vehicles in the city, road capacity becoming insufficient.There are only two over bridges in the city. As parts of the city frequently get inundated in storm floods during the monsoon season mainly in the months of June,

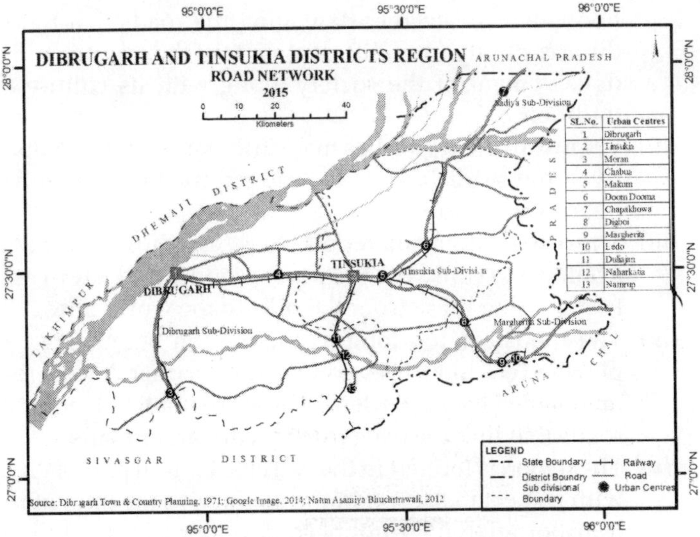

Fig 13.2: Dibrugarh and Tinsukia District Region Road Nework, 2015.

July and August, city roads are being constructed with concrete. It is said that Dibrugarh is the first city in the North-east to have introduced city bus service more than half a century back some time in 1956. Up till the mid-1990s, City Buses and cycle rickshaws were the main modes of urban transport. Later, 'Vikram', 'Minidor', 'Champion' etc. three-wheelers were also introduced as means of urban transport and replace the city bus service.

Dibrugarh has three bus terminuses — ASTC bus terminus at Chowkidingee, Muralidhar Jalan Bus Terminus at Malakhubasa and AMCH bus terminus at Borbari near the AMCH gate. For inter-city journey, luxury buses including Volvo buses plying between important destinations within Assam and other neighbouring states at frequent interval are available.

Problems

There are few reasons for backwardness of the region in respect of road transport. These are:

(i) The rainfall being very heavy in summer, there occurs

floods which severely damaging the roads that hinders the accesibility from different parts of the country and the development of the society along with its culture and tradition.

(ii) Because of the region has many tributaries of Brahmaputra river, the roads have to be bridged frequently involving huge expenditure.

(iii) The road-building materials like stone chips are not easily available to construct the road, requiring a heavy cost in transporting these from the hills to the works sites.

(iv) The roads are too long in this region because tea, oil, plywood etc. industries grew-up here earlier. A set of roads maintained by tea gardens. These are small but motorable roads that link the tea gardens with larger roads.

(v) The region is located in the extreme eastern part of Assam with respect to the rest of the country, so the lines of road transportation have not been developed to the desirable extent and hence the international market network which combines different tribes and traditions is suffering a lot.

(vi) Apart from tea and oil industries, there is practically no large industry in the region. Therefore, a well-knit network of road transport has not been laid so far.

(vii) State Governments and Administrator of the region is not financially well off to afford sufficient expenditure to construct and maintain a well-knit transport system. Hence construction of new roads is rare and maintenance of existing roads is poor.

Prospects

National Highway Authority of India is proposing to make Dibrugarh, a NHAI hub after the completion of Bogibeel bridge. Dibrugarh is connected to the rest of India by NH-37. After the completion of Bogibeel bridge, it will be connected to northern bank of the Brahmaputra by NH-52(B). A new bypass is being constructed to ease traffic movement within the city area near Jameera. A new flyover has also come-up along the NH-37 over the newly developed Dibrugarh-Moran rail line near Dibrugarh University.

In respect of number of motor vehicles on road is the only

region with about 1,78,583 vehicles in 2011-2012 comes next only to Kamrup having the highest number of motor vehicles on road in the state. The number of buses, motor cars, jeeps, taxi cabs and motor cycles and scooters, on road in the districts region as recorded in 2011-2012 are 766; 47904; 621; 2431 and 1,03,634 respectively. Opening of border by Stilwell highway with Myanmar will have great impact on the development of the region.

Conclusion

There is an abnormal gap between the existing and expected length and efficiency of the roads. The fact is that the density of roads per hundred sq.km. of area and the road length per lakh of population are lower than the national average in the region. Greater emphasis is needed to expand the surfaced roads in the rural areas. Besides this expansion of road network the demands of the people will not fulfil unless it is upgraded through some qualitative improvement. A large number of villages are yet to be connected by all-weathered roads. Whatever road extension programme is going on in the state, it is seen that most of them are guided by political considerations rather than the needs of the public demand. As a result, the systematic mobilization of resources, better co-ordination and management including agriculture should be needed for future growth and development of the Dibrugarh and Tinsukia Districts Region.

References

Bhagabati, A.K.; Bora, A.K.; Kar, B.K. (ed.) (2001) *Geography of Assam*, Ne. Delhi: Rajesh Publications, p. 242.

Borooah, G.L. (1985) *Population Geography of Assam: A Study of Dibrugarh District*, New Delhi: Mittal Publications.

Dutta & Dutta (1901, Rep. 1976) *Assam District Gazetteers*, Guwahati, Assam, p. 306.

Goswami, P.C. (1994) *The Economic Development of Assam*, Ludhiana, New Delhi: Kalyani Publishers.

Government of Assam, *Statistical Hand Book*, Assam (1961, 1971, 2012): Directorate of Economics and Statistics, Guwahati, Assam.

Taher and Ahmed (2001), *Geography of N.E. India*, Guwahati: Mani Manik Prakash.

14

Indian Export and Prospect of Re-opening Stilwell Road: An Analysis

Jagadish Nath and *Mokhalles M. Mehdi*

Introduction

Globalisation has increased the importance of trade and opportunities for businesses. Countries have started forming regional trade blocs and economic integration with neighbouring nations. Cross-border trade has become the medium of current wave of globalisation. The Government of India announced the "Look East Policy" in the early 1990 to pave the way for international trade between North-east states (NE states) of India and the South-East Asian countries. India has focussed on cooperation between countries to maintain the trade practices among NE states and neighbouring regions. The North-east have been located at a crossroads between three major economies – East Asia, South Asia and South-east Asia. The North Eastern Region (NER) comprises of eight states includes Arunachal Pradesh, Assam, Manipur, Meghalaya, Mizoram, Nagaland, Sikkim and Tripura. NER shares two per cent of the border with the mainland of the country and more than 98 per cent is linked with international borders with BBCMN (Bangladesh, Bhutan, China, Myanmar and Nepal) countries. There could be trade and economic connectivity between India's neighbours with the ASEAN countries namely, Myanmar and Thailand. The economic connection can be better with good infrastructure. The Bay of

Bengal Initiative for Multi-Sectoral Technical and Economic Cooperation (BIMSTEC) have been a good framework for regional integration highlighting the easier way of connecting countries within a regional framework.

NER is the "Gateway to South East Asia" and Look East Policy is viewed especially favourable for boosting trade between North East India and the South East Asian Nations. It is a practice that countries normally conduct the trade with neighbouring countries who share geographical boundary. Trade history have witnessed that the neighbouring countries of NER had been the traditional markets for a variety of goods and services produced by the North-east before its traditional link were disrupted by the partition in 1947 (Mukherjee, 1992). Therefore, the present study is an attempt to investigate the Indian export to neighbouring countries of North-east India and critically evaluate infrastructure development in the region.

Objectives

1. To examine the export share of India with ASEAN, South Asia and NE Asia region specific to neighbouring countries of North-east India.
2. To analyse the issues and challenges of revival of Stilwell road connectivity with India and South-east Asia.
3. To provide the measures for improvement of business in North-east India.

Methodology

The present study is based on secondary informations. The secondary information has been collected from the government websites such as, Ministry of Commerce and Industry, trade portal of India, export promotion organizations and research articles to satisfy the objectives.

Findings and Analysis

The North-east region (NER) of India consists of eight states namely, Assam, Arunachal Pradesh, Manipur, Mizoram, Meghalaya, Tripura, Nagaland, and Sikkim. North-east region of

India has shared geographical boundary with five neighbouring countries viz., Bangladesh, Bhutan, China, Myanmar, and Nepal (BBCMN).

Region-wise, Export share of India

Table 14.1 points out that India has been exporting to several geographic regions across the world. The countries of eastern region belong to Association of South-East Asian Nations (ASEAN), North-east Asia (NE Asia) and South Asia. Data indicates that Indian exports to different geographic region was constantly increased during the financial year (FY) 2009-10 to FY 2011-12. However, it has been seen that Indian exports were marginally dropped during the FY 2012-13 and shown an export growth of 5 per cent in the FY 2013-14 compared to the last financial year.

Moreover, Indian export to South Asia was constantly increased during the five fiscal years from FY 2009-10 to FY 2013-14. Similarly, Indian export to ASEAN and NE Asia have seen an increased rate for three years followed by a negative growth in the FY 2012-13, and subsequently increased in FY 2013-14. Indian export have seen the growth of 4 per cent and 16 per cent in NE Asia and South Asia respectively for the period of FY 2012-13 to FY 2013-14.

In the FY 2013-14, India has the highest export to European Union countries and lowest to Central African Republic (CARs) countries. Although, Indian export was highest to the European Free Trade Association (EFTA) followed by Other South African Countries and Central Africa at the growth rate of 48 per cent, 23 per cent and 17 per cent, respectively for the period of FY 2012-13 to FY 2013-14.

Export Share of India with BBCMN Countries

Table 14.2 reveals that India has been continuously exporting to BBCMN countries as well as other regions of the world (refer Table 14.1). Data of FY 2013-14 highlights that India's share of export to BBCMN countries was more than eight per cent of global exports.

Data indicated during the FY 2009-10 to FY 2011-12 Indian exports to these five countries was continually increased. However, FY 2012-13 indicates about the overall decline in Indian export to BBCMN countries. In FY 2012-13 to FY 2013-14 export growth rate

Table 14.1: India's Export with World—Region-wise

Values in US$ Million

Sl. No.	Regions	2009-10	2010-11	2011-12	2012-13	2013-14	%Growth of 2012-13 to 2013-14
1.	EU Countries	36,028	46,039	52,556	50,422	51,581	2
2.	European Free Trade Association (EFTA)	835	954	1,464	1,378	2,046	**48**
3.	Other European Countries	1,660	2,893	3,731	4,202	4,644	11
4.	South African Customs Union (SACU)	2,178	4,107	4,890	5,281	5,395	2
5.	Other South African Countries	1,130	1,512	1,328	1,886	2,329	**23**
6.	West Africa	3,137	4,297	6,460	6,523	6,993	7
7.	Central Africa	350	465	708	931	1,092	**17**
8.	East Africa	3,512	5,347	6,594	8,839	9,975	13
9.	North Africa	3,125	3,986	4,694	5,682	5,442	–4
10.	North America	21,254	27,554	38,163	39,820	43,407	9
11.	Latin America	5,614	9,324	12,277	13,518	10,792	–20
12.	East Asia (Oceania)	1,691	2,506	2,817	2,733	2,683	–2
13.	**ASEAN**	**18,114**	**25,628**	**36,744**	**33,008**	**33,134**	**0**
14.	West Asia—GCC	30,480	42,472	45,360	51,054	48,221	–6
15.	Other West Asia	5,749	7,760	9,536	11,356	13,067	15
16.	**NE Asia**	**28,905**	**35,960**	**45,303**	**39,388**	**40,773**	**4**
17.	**South Asia**	**8,391**	**11,657**	**13,296**	**15,111**	**17,504**	**16**
18.	CARs Countries	269	303	430	551	538	–2
19.	Other CIS Countries	1,418	2,379	2,630	3,131	2,954	–6
20.	Unspecified	4,910	14,673	16,983	5,587	11,836	112
	India's Total Export	**178,751**	**249,816**	**305,964**	**300,401**	**314,405**	**5**

Source: Author, compiled data from tradeportalofindia.com

Table 14.2: India's Export to BBCMN Countries

(Values in US$ Million)

Sl. No.	Country	2009-10	2010-11	2011-12	2012-13	2013-14	%Growth of 2012-13 to 2013-14
1.	Bangladesh PR	2,433.77	3,242.90	3,789.20	5,144.99	6,166.93	19.86
2.	Bhutan	118.86	176.03	229.86	233.22	355.6	52.47
3.	China PRP	11,617.88	14,168.86	18,076.55	13,534.88	14,824.36	9.53
4.	Myanmar	207.97	320.62	545.38	544.66	787.01	44.5
5.	Nepal	1,533.31	2,168.06	2,721.57	3,088.84	3,592.30	16.3
	Total	**15,911.79**	**20,076.48**	**25,362.54**	**22,546.58**	**25,726.19**	**14.1**
India's Total		**178,751.43**	**249,815.55**	**305,963.92**	**300,400.58**	**314,405.30**	**4.66**
%Share		**8.9016**	**8.0365**	**8.2894**	**7.5055**	**8.1825**	

Source: Author, compiled data from tradeportalofindia.com

of India with these countries was 14.1 per cent. Out of these five countries, Indian export was highest to Bhutan followed by Myanmar, and slow growth with China during the FY 2012-13 to FY 2013-14.

Data indicated that Indian export to Bangladesh, Bhutan, and Nepal was constantly increased during the FY 2009-10 to FY 2013-14. Indian export to China and Myanmar was constantly increased from FY 2009-10 to FY 2011-2012, followed by a decline growth during FY 2012-13 and subsequently increased in FY 2013-14.

Overview of Stilwell Road Map

The Stilwell road is about 1,726 kilometres long from India to China *via* Myanmar (refer to Fig. 14.1).The major portion (1033 kms) of the road runs through the territory of Myanmar, as the kingpin of the connectivity between India and China. The road between Myktina, capital of Kachin state and Pangsau pass up to the Indian border. Myanmar has agreed to upgrade the connectivity through

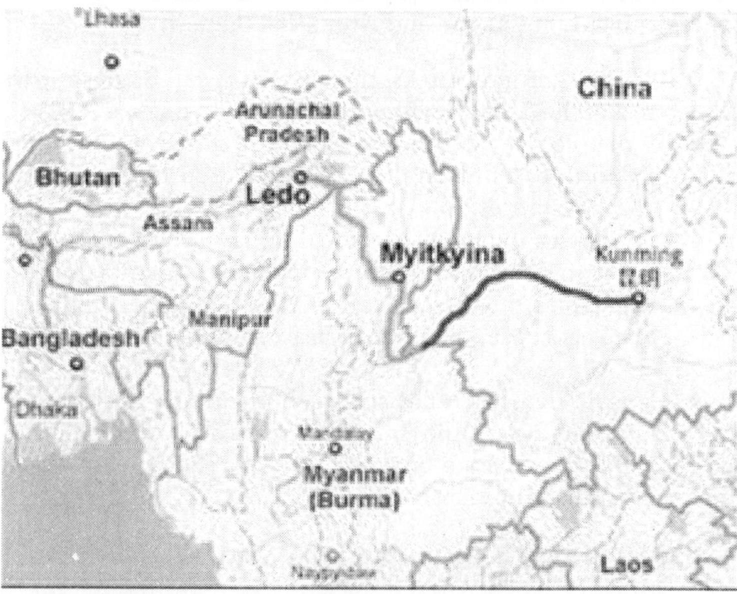

Source: Puyam, G.S. (2015), *"Opening Stilwell Road: Opportunities and Challenges"*, CAPS INFOCUS March: 1-5.

Fig. 14.1: Stillwell Road

Chinese investment. However, China has the road of 632 km up to Tanai of Kachin district and China has already constructed the road. India has not yet upgraded the connectivity of 61 kms of road from Ledo (India) to Indo-Myanmar border due to security concerns.

Opportunities

The Stilwell road will provide an opportunity for economic growth and cultural cooperation between India and South East Asia nations. Considering the example, a cargo movement from India to China took more than three weeks from Kolkata to China by a sea-route. However, the duration of transportation can be reduced to less than a week if cargo transportation will be done by Stilwell road to Yunnan in China. Thus, the strategic investment on building the road infrastructure will focus on achieving India's Look East Policy. Moreover, it will enhance people to people interaction through tourism, cultural programmes, trade, and commerce. Also, will increase the bond between people of the connected countries. The Stilwell road could bring some of the key benefits for North-east India including Assam and Arunachal Pradesh are:

- The border trade of Assam and Arunachal Pradesh with rest of the country will increase.
- Indian companies from mainland will start investing more in North-east India for the growth of export to South East Asian countries.
- North-east India will become the regional hub for Indian automobile components, pharmaceuticals, fruits, vegetables, textiles or cotton yarn, electronic gadgets, blankets, consumer durables, teak, gold and semi-precious stones.
- Unexplored potentials such as horticulture, floriculture, biodiversity and tourism will increase in North-east India.
- Cultural exchange between India and South-east Asian countries would be more.

Challenges in Reopening the Stilwell Road

There are several challenges that the government will face in reopening the Stilwell road for India, firstly, Arunachal Pradesh will be flooded with Chinese goods and secondly China claims Arunachal Pradesh as its own territory. It would be better for India to connect and establish a proper infrastructure between mainland of India and Arunachal Pradesh instead of developing a road link with China, considering it as a priority. Without proper road infrastructure between mainland of India and Arunachal Pradesh, entire North-east India will be flooded with cheap Chinese goods and consumer durables. Investment in the infrastructure sector in remote regions always is counted as a strategic project. Therefore, it can be concluded that instead of developing the economy of North-east India, revival of Stilwell road may undermine the local economy. Moreover, opening of this road provides an access from China to Arunachal Pradesh through Myanmar, which is a concern for border security because these regions has influenced by insurgents and huge ethnic divisional conflicts.

Conclusion and Suggestions

The infrastructure development of North-east region is vital for the promotion of business in India with South-east Asian countries. The key reason for underdevelopment of North-east region includes geographic positioning, and economic factors. However, several initiatives have been identified such as strengthening the existing cross-border trade link with ASEAN countries, and increase the connection of cross-border trade with other NER state capitals and business hubs. It is true that in NER physical infrastructure is the key concern such as majority portion of the road are unprepared and needs upgradation to build the transport connectivity. Secondly, upgradation of railway lines from meter gauge to broad gauge lines for smooth transportation of goods should be extended across the North-east. Lastly, it is necessary to have air connectivity among these states. Hence, non-functioning airports have to improve immediately for providing transport linkages. The air connectivity would increase international trade with ASEAN countries.

At the same time, policy-makers should take some key measures for the improvement of business environment in NER region. These includes (i) promotion of local products; (ii) adequate logistic support and communication; (iii) travel distance minimization of goods movement in export by proper arrangement with Bangladesh; and (iv) involvement of export promotional organizations to increase the business in the region. In addition, government have to provide adequate entrepreneurship training, financial and administrative support tc local producers for manufacturing as well as marketing of the products.

References

Jyoti (2013): "India's Look East Policy: In its Second Phase", *Global Journal of Pharmaceutical Sciences and Education*, 2(1):1-14.

Mukherjee, B.N. (1992): *External Trade of Early North-East India* (New Delhi: Har-Anand Publication).

Puyam, G.S. (2015): "Opening Stilwell Road: Opportunities and Challenges", *CAPS INFOCUS*, March: 1-5.

Strachan, A.L., Kang, H.K., and Sinha, T. (2009): "India's Look East Policy: A Critical Assessment", *IPCS Special Report*, October: 1-10.

15

Exploring the History of the Road Communication in Manipur During the Nineteenth and the Twentieth Century

Console Zamreinao Shimrei

Introduction

Manipur is situated at the eastern frontier of India. The area of Manipur state is 22,327 sq. km. since the transfer of Kabaw Valley to Burma in 1834. It is bounded in the north by Nagaland, in the east and the south by Burma, in the south-west by Mizoram and in the west by Assam. Manipur is geographically divided into the hills and the Valley. The Imphal Valley is 790 metres above sea-level and surrounded by the hills in all sides. This geographic division is a determinant factor in the socio-political and historical development of the state. The Imphal Valley inhabited by the Meiteis is one-tenth of the total area of the state. Ninety per cent of the geographical area belongs to hill tribes people of Manipur. The hills area is mainly inhabited by the Nagas and the Kukis tribes respectively.[1]

This chapter attempts to explore the history of the Road Communication in Manipur during the nineteenth century and twentieth century. Road communication in Manipur deserves the maximum attention from administrative, military, political, education, economic, and social points of view. Transport and communication play an important role in bringing integration of the heterogeneous social and political units on the one hand and in breaking the age old rigid economy system of Manipur on the other. Historically speaking, Manipur is one of the most backward states so far as the road communication system is concerned. There were various factors which attributed backwardness of Manipur state specially road communication in the nineteenth and twentieth century. The first reason was that the major part of the state is covered by a number of hill ranges in all sides. It was difficult to access from one village to another village due to topographical conditions. Secondly, it is believed that the Valley of Manipur was quite safe from the security and defence point of view where the Imphal Valley was ruled by the *Rajas* since ancient times. Hence, they could maintain a separate political and administrative unit of their own without the interference of outsiders.[2] Another reason was that the hills tribes like the Tangkhul Nagas of Manipur settled on the top of steep hills for security point of view.[3] Since the Tangkhul Nagas were known for headhunter even in the eighteenth century. Moreover, the sources of revenue of the state were extremely limited for meeting the expenses on the construction of roads and bridges. Although there were a few surface-routes, yet the state remained mostly a footpath for a long time. In the absence of road communication, water transport played the most important role in the transport and communication system of Manipur. Transport and road communication helped people to move out from one place to another and enabled to export and import goods and materials for some specific purpose to meet the requirement of the people. In fact, the oldest form of transport relied on human labour. Animals were then used for transportation of goods and services. Invention of wheel, for making pottery, was one of the greatest contributions of the Neolithic man to revolutionize the means of transportation and road communication with the use of wheeled cart drawn by men. It made transportation much easier and simpler. As a result, animals were used to draw it, such as bullock cart and chariots. The first wheeled military vehicles were developed around 2500 B.C. Since then, significant amounts of resources was spent by rulers and their builders to construct and maintain communication routes in the form of roads.[4]

Another means of transport communication used in ancient times was raft or floating log of wood or bamboo. It carried loads along river downstream. Moreover, sailing vessels greatly enlarged the scope of human movement and had empowered man to communicate with one another and practiced trade over long distances.[5] In view of the fact that, the geographical location of the state could be assumed that rivers of the Imphal Valley were used for movement of people and goods in ancient times by using local made boats. However, water transport was mainly practiced in the Imphal and Iril rivers and in the Loktak Lake. Another important means of transportation goods was *Kangkot* (*Kangkot* is a local made cart without wheel used widely in Manipur during the early period) drawn by animals such as bullock or buffaloes. Indeed, it was only after the British domination over the kings of Manipur (since 1826) that the bullock cart was used for the transportation of goods. However, bullock cart is used till today.

Before the Second World War, the transport and communication systems of Manipur were extremely poor. It was very difficult to reach many interior villages particularly the hill villages. Road communication in Manipur began around 1830s to 1840s with the advent of the British Empire in the princely state of Manipur. In fact, construction of roads in the hill areas became an important feature to connect Manipur with the neighbouring states. Construction of Old Cachar Road was taken-up in the expense of the British in 1837 and was completed in 1844 into a cart road. Later on, the construction and maintenance of this road was undertaken by the then king of Manipur (*Annual Administration Report,* 1844). Interestingly, in the early part of 1891, the construction of Imphal – Dimapur road was completed by the British to invade Manipur. In the same year, the English people invaded the territory of Manipur and the road was opened for commercial purpose in February 1894.[6]

It could be mentioned that till the last quarter of the 19th century, there was hardly any motorable road and hence no wheeled cart could ply in this region. Therefore, porters and cattles were commonly used as the chief carriers. In the absence of any wheeled cart that *Kangpot*, a kind of cart which did not have wheels, was used by the people of Manipur for carrying loads. Horses and elephants were other important means of transport and communication especially used by the *Rajas* for their travels and for carrying their messages. After the opening of cart road in 1881 that bullock carts could be seen in the state and that to have a bullock cart by a family began to be considered as one of the most important assets of a household. Besides, bicycles, motor-cycles and lorries were brought in the state to ply by 1920s. However, the road communication system of Manipur could be divided into two categories: Inter-village roads and Inter-state roads respectively.

Inter-Village Roads

Inter-village roads were those which connected the villages of the state both in the hills and in the valley. In this connection, before 1891, there were a few important roads constructed by the *Rajas* using the services of the *Lallup* labour. During the reign of Maharaja Chandra Kirti, a narrow road from Imphal leading to the western ranges of the state was constructed in 1873.[7] Nevertheless, most of the inter-village roads constructed by using *Lallup* labour were confined to the surrounding area of palaces of various *Rajas* at different periods. Thus, the most popular inter-village roads in the valley till late 19th century were the river banks. On the other hand, most of the inter-village routes were generally made in the form of footpaths.[8] As a result, some of these footpaths were not actually constructed but they became so after repeated use of them by the villagers.

Towards the end of the 19th century, the road connection was made from Imphal to Moirang village which was made Jeepable road. Another cart road across the Ngarial hills leading to Yairipok to the east of Imphal was also started in 1905.[9] Further, another important route connecting the Imphal Valley with the Ukhrul village on the north-eastern hill ranges was also proposed for construction of road in 1909 and it was completed in 1922.[10] In 1920, the Government of British India drastically reduced the tribute, which was payable to them by the *Raja* of Manipur from Rs. 50,000 to Rs. 5,000 per annum for a period of ten years so that the state might be able to spend the balance for the improvement of inter-village roads, especially in the hills area. Thus, the first quarter of the 20th century was a turning point so far as development of inter-village roads communication in the state is concerned. As a result, a new road with a length of 19 kilometres from Tuinem to Karong was constructed. Besides, a new diversion road with a length of 53 kilometres on the Imphal-Ukhrul road and an extension of 13 kilometres of bridle path from Ukhrul to its interior parts of the villagers were also constructed by 1925. In addition to these, other roads, namely, the Onamlong-Tamenglong, the Nungba-Tamenglong and the Phellong-Tamenglong all in the north-western part of the state were found to have been completed.[11] The Mayang-Imphal road was also constructed into a Jeepable road up to Sekmaijin, a village about 30 kilometres to the south of Imphal. In 1927, the Imphal to Moirang road was extended up to Churachandpur, 61 kilometres to the south of Imphal. Similarly, extension of another bridle path was constructed from Hundung near Ukhrul village to Kasom village (extreme North-east boarder of Manipur) in 1929-30.[12] Hence, the state had to spend a sum of Rs. 50,848 and Rs. 23,719 for the maintenance of all these roads in the valley and the hills respectively in 1924-25. Nevertheless, all these roads had to be improved and extended further as most of the villages of the state in general and the villages in the hills in particular were not yet connected properly with one another. The Manipur Government taken up various steps for construction of more numbers of roads in the state. As a result, in 1950, the Kangchup road, the Sagolband road and the Pukhao road were constructed in the Imphal valley. The Tidim road connected the southern and the south-western hills; the Imphal-Yairipok road connected the eastern hills and the Imphal-Ukhrul road connected the north-eastern hills etc.[13] That, Imphal to Kongkang Thanah (Tangkhul village) *via* Shangshak, Kumnou, Kamjong, Mavailung villages etc. was very important road during the Second World War. Mostly the Japanese soldiers passed through this route and they captured Shangshak village, Harvakhangai (Khangkhui village in Ukhrul), Moirang at Imphal and they even captured Kohima through the help of this route. Thus, the total expenditure of the state for the maintenance of these roads both in the valley and the hills were estimated at Rs. 1,82,920 and Rs. 62,393 respectively. The total estimated length of the inter-village road of the state was at about 356 kilometres in 1951.[14]

Motorable road was constructed from Nungshong Khullen to Nungshong Khullen BRTF road Junction. Initially, this road was constructed by the Veikhang area like Nungshong Khullen, Nungshong Khunou, Choither, Khangkhui Khullen and Khangkhui Khunou villages, to enable to sale their villagers' products like potatoes, maize, sesame etc. at Sena Keithei (the main bazaar at Imphal). Today, this road was constructed from Nungshong Khullen BRTF road Junction to Khangkhui Khunou village under

Pradhan Mantri Gram Sadak Yojana (PMGSY) from one side and National Highway No. 150 to Choithar village on the other side under the same programme. However, this road from Choither village to Khangkhui Khanou village is yet to connect.

Inter-State Roads Communication

Manipur is connected with Cachar in Assam, Sylhet etc. towards the west and Burma, China etc., towards the east by foot and bridle paths. Saletore (*Early Indian Economic History*) clearly mentioned that there were three important routes connecting India and China of which the shortest one passed through Manipur.[15] It was also mentioned that three routes connected Manipur with Cachar in Assam. The first route was the Aquee route and the second one was the Kala Naga route which start from Bishenpur in Manipur to Baskandi in Cachar district of Assam and passed through many hill ranges covered with dense forests. The third one was the Khongjai route which start from Torbung village in the south western part of the state and proceeded *via* the western bank of the Barak leading to Baskandi. Another important bridle path which connected Manipur with Assam by crossing the western ranges of Manipur at the Tholang village and then leading to the north-west direction up to the Dansiri river and proceeded to Jorhat. Interestingly, Maharaja Gambhir Singh and Captain Gordon attempted to explore another road communication in 1833 which could be the shortest one to reach Assam from Manipir.[16]

Manipur had well road communication with Burma since the ancient times. The most important two routes were the Heirok route and the Imole route, both of which led to the Kabaw valley in Burma after crossing the Maring hills of Manipur. Another important road communication which led to the Burma was connected from Ukhrul, the eastern most part of Manipur. Thus, the native people of Manipur could reach the western and eastern states through these routes. Notwithstanding, the main feature of these routes was that they had to pass through dense forests, streams and rivers. During the raining season, it was very difficult to cross them. According to Johnstone, some sort of suspension bridges were found to have been constructed over the bigger rivers like the Barak and Erang rivers in the 19th century.[17] With the passage of time, some of these foot and bridle paths had been constructed in broader roads. They were the Manipur-Cachar road, the Tamu-Kohima-road etc. thus, these were only two main road-links between the states of Manipur and other parts of India.

(i) Manipur-Cachar Road

The British helped Maharaja Gambhir Singh to drive out the Burmese from the territory of Manipur in 1826. After that a new era of understanding and co-operation between Manipur and the British East India Company was established. The development of roads communication with the provinces of the British East India Company became very important especially for the trade and commerce, integration and defence of the North Eastern Frontier of British India from Burmese invaders. In 1827, the British East India Company persuaded Raja Govind Chandra of Cachar to construct road between Sylhet and Manipur *via* Cachar.[18] However, Govind Chandra was assassinated in 1830 before construction of this road and his kingdom was annexed to British India in 1832 leaving the responsibility of the construction of the proposed road with the British East India Company.[19]

Subsequently, in 1833, Maharaja Gambhir Singh signed a treaty with the British India Government. In accordance with the treaty, Manipur was obliged to provide all possible assistance to the British officers especially for the construction of road between Manipur and Cachar in Assam. Thus, the construction of the road started in 1837 at the joint expenses of the Government of British India and Manipur under the supervision of Captain Guthire and it was completed in 1842.[20] The length of the portion of the road from Imphal *via* Bishenpur, Khoupum, Nungba and Jiribam at the border between Manipur and the old Cachar was about 180 kilometres. The road communication systems in the state improved to a great extent after achievement of independence and taking up successive Five Year Development Plans and Communality Development project. Thus, a New Cachar Road (NH-53) having a length of about 250 km. stretching from Imphal to Cachar district of Assam passing through Noney, Nungba and Jiribam Sub-division etc. was constructed by the Border Roads Task Force.[21] Thus, more than eight new bridges were constructed near Jirighat in 1906 and the important bridges were furnished with iron girders.[22] However, with the construction of the Manipur-Kohima road after 1895, the importance of the Manipur-Cachar road was decreased and hence it received comparatively less attention. But when the Manipur-Kohima road was delinked temporarily by the Japanese forces during the Second World War, some steps were taken-up again to improve the Manipur-Cachar road with the help of the military authority of India.

(ii) Tamu-Manipur-Kohima-Dimapur Road

The Tamu-Manipur-Kohima-Dimapur road which had a total length of about 215 kilometres was a narrow, winding, rough road with many steep grades and sharp curves at several points. Till the late twentieth century small carts pulled by men and light vehicles were used on this road for transport of rice and other essential commodities from Dimapur.[23] This National Highway-39 starts from Tamu in Burma *via* Imphal, Senapati, Maram, Mao gate, Kohima and Dimapur. After the Anglo-Burmese War (1824-25), the officers of the British East India Company tried to convince Maharaja Gambhir Singh of the importance of this road which had to pass through the Naga Hills. Therefore, in 1832, Pemberton and Captain Jenkins escorted by 700 Manipuri troops and 800 coolies attempted to pass through Naga Hills to access the possibility of construction of this road. In this connection, it may be mentioned that since there was no good road communication from Imphal to Mao villages so proposal was made for construction of cart-road for the first time in the 3rd week of January, 1880 and its construction was completed in 1881.[24] Hence, this road was found to be very beneficial to British India, especially during the Anglo-Burmese War of 1885.

During the reign of Maharaja Chandra Kirti, Imphal to Burma border road which was the extension of Imphal-Mao cart road was constructed. But considering the Imphal-Mao road which led to Kohima as more important, higher priority was given to it for further improvement. Hence, the importance of this cart road increased considerably after the British occupation of Manipur in 1891. Thus, the government of British India immediately put enormous efforts not only to improve this road but also to extend road communication beyond the boundary of Manipur that was up to Kohima. They tried to complete it in 1895. Since the road was still unmetalled, heavy rain caused damages very often on the road which demanded a huge amount of money for improvement and repairing. Therefore, metalling of the bad patches on the Manipur-Kohima cart road started in the beginning of the 20th century, while the extensive metalling work of the road took place only after 1907. After that the authority of British India explored road communication up to Tamu in Burma on the other.[25] As a result, improvement of road connectivity between the Dimapur and Tamu began to meet the need of the military design of the British India.

In spite of the vigorous counter offensive measures made by the British troops, the Japanese forces could successfully stop the supply of the British warrior through Manipur-Dimapur road and they had blown-up one of the vital bridges on this route. In return, the British Indian military forces resorted to repair and improved the Cachar road while repairing of the bridge on the Manipur-Kohima road was going on. Thus, after the Second World War, particularly in 1946, the military authority stopped the responsibility of the maintenance of the road from Mao to Tamu by leaving it to be a tarmac road. Later on, this road became the most important inter-state road connecting Manipur with the Indian states on the one hand and with Burma on the other hand. In the meanwhile, the Government of India decided to declare the inter-state roads which included the Tamu-Manipur-Kohima road as National Highway No. 39 in 1947.[26] Similarly, the Manipur State Darbar also endorsed the decision of the Government of British India on the 30th April, 1947.[27]

The Government of British India insisted on the state to contribute larger share of the estimated total cost. Thus, the Manipur State contributed Rs. 30,000 annually towards the maintenance of the road since 1894-95. The average total estimated cost for maintaining the Imphal-Mao road for three years from 1895 to 1898 was Rs. 54,603.[28] But the Government of British India insisted on further increase of the state share when the road became metalled. However, Maharaja Churachandra Singh as well as the people of Manipur resisted to the demand as they considered the construction of the road did not bring any worth benefit to the state. Rather, they suffered from occasional food shortage due to overexportation of rice from the state through this road communication facility. However, this road was the most important and became the lifeline of Manipur.

In 1950, the total length of surface road of Manipur was estimated as 651 kilometres.

Important Surfaced Road in Manipur as found in 1951 are given below:

Sl. No.	Name of the Road	From	To	Length in (km)
1.	Kangchup Road	Imphal	Kangchup	14
2.	Sagolband	Imphal	Khumbong	14
3.	Pukhao	Imphal	Sagolmang	20
4.	Tidim Road	Imphal	Churachandpur	61
5.	Mayangimphal	Imphal	Sugnu	64
6.	Yairipok	Imphal	Ningel	40
7.	Ukhrul Road	Imphal	Ukhrul	86
8.	Burma	Imphal	Moreh	111
9.	Kohima Road	Imphal	Mao	107
10.	Cachar Road	Bishenpur	Jirighat	135

Changes

The state has 8,648 km of roads both metalled and unmetalled. The length of roads consists of State Highway 668 km., National Highways 967 km., District roads 1,977 km and village roads 5,036 km.

National Highway No. 39: It is known as Indo-Burma Road or Dimapur Road. Today, this National Highway name was chang into National Highway No. 2 instead of 39. It is the most important National Highway of Manipur state. Its total length is 215 kms. The nearest rail station at Dimapur in Nagaland could be reached only through this road.

National Highway No. 53: It is also known as New Cachar Road. This also changes into National Highway No. 37 instead of National Highway No. 53. It connects Imphal with Jirighat in Cachar district of Assam. Its total length is 224 kms.

National Highway No. 150: It was another important National Highway running through Manipur. It is the road connecting Jessami-Ukhrul-Imphal-Churachandpur-Tipaimukh with Aizwal of Mizoram. Manipur's share of the proposed NH-150 would be 523 kilometres long.[29]

Thus, the state of Manipur is well depended on road communication for its transportation. The state is included in the railway map of India with the opening of a rail-head at Jiribam in

May 1990. A railway line of 50 km connects it with Silichar in Assam railway station. It is 255 km from Imphal. Dimapur, 215 km from Imphal is the nearest rail-head. The second railway line will be between Karong in Senapati District and Diphu in Assam through Dhansiri. The foundation stone was laid on November 17, 1998. Another line to be connected with Silichar station with Toupul in Manipur was inaugurated in November 2004.

Notes

1. Kabui, Gangmumei, 2011, *History of Manipur Volume One Pre-Colonial Period*, New Delhi: National Publishing House, pp. 7-8.
2. Ali, Md. Ashraf, 1988, "*Economic History of Manipur (1801–1951)*," Unpublished Ph.D. Thesis, Imphal: Manipur University, Canchipur, p. 128.
3. Saha, N., 1978, *Customs and Economy—A Study of Shifting Cultivation in North-East India*, In Dubey, S.M. (ed.): *North-East India—A Sociological Study*, New Delhi: Concept Publishing Company, p. 337.
4. Jotin Khisty and Kent Lall, B., 2002, *Transportation Engineering—An Introduction*, New Delhi: Prentice-Hall, p. 2.
5. Senior, M.L., and White, H.P., 1983, *Transport Geography*, London: Longman Group Limited, p. 10.
6. *Annual Administration Report of Manipur*, Shillong : Government Press, 1894-95, p. 3.
7. *Annual Administration Report of Manipur*, 1873-74.
8. McCulloh, M., 1980, *Account of Valley of Manipur*, Delhi: Gian Books, pp. 43-44.
9. Political Agent's Tour Diary of 31st March, 1897 and 6th to 31st March, 1907.
10. Political Agent File No. 195. Letter from the Secretary to the Government of East Bengal & Assam to the Political Agent, Manipur, No. 5603 dated 6-10-1909.
11. *Annual Administration Reports of Manipur*, 1920-21 to 1925-26.
12. *Annual Administration Report of Manipur*, 1929-30.
13. Government of Manipur, Home and Development Departments, Letter from the Secretary to the Government of Manipur, to the Chairman, Town Fund, Imphal No. Budget/1/50/5914. H.D. dated 30-8-1950.
14. *National Council of Applied Economic Research* Techno-Economic Survey of Manipur, Imphal: Manipur Administration, 1961, p. 51.
15. Saletore, R.N., 1973, *Early Indian Economic History*, Bombay: N.M. Tripathi Private Limited, p. 390.

16. Singh, R.K. Jhaljit, 1965, *A Short History of Manipur,* Imphal: O.K. Store, p. 3.
17. Johnstone, J., 1988, *Manipur and Naga Hills,* Delhi: Cultural Publishing House, p. 97.
18. Bhattacharjee, J.B., 1977, *Cachar Under British Rule in North-East India,* Delhi: Radiant Publishers, p. 45.
19. *Foreign Department Political Consultation—A Proceeding No. 17.* Dated 17-9-1832.
20. *Annual Administration Report of Manipur,* 1875-76.
21. Singh, L. Udhob, 1989, *A Brief History of Agricultural Development in Manipur (1946-1983),* Imphal: R.B. Printing House, pp. 24-25.
22. *Annual Administration Report of Manipur,* 1906-07.
23. Singh, L.Udhob, *op. cit.,* p. 24.
24. Singh, R.K. Jhaljit, *op. cit.,* p. 206.
25. *Political Agent's Tour Diaries of 9th May, 1895 and 27th May, 1906.*
26. *Political Agency File Subject: National Highway No. 39.* Letter from the Government of India War Transport Department to Government of Assam, P.W.D. No. PL-13 (36) dt. 25-7-1946.
27. *Political Agency File, D.R.* No. 11 of 30-4-1947, Memo No. 1305 GA/506 date 24-5-1947.
28. *Political Agency File, No. 163,* Letter from E.J. Mitchel, Esq. C.E., Executive Engineer, Manipur division to Superintendent Engineer, Assam Division dated 7-2-1898.
29. Gupta's, R., 2014, *Manipur General Knowledge,* New Delhi: Ramesh Publishing House, pp. 168-170.

Section D
Road and Its Impact

16

The Making of National Highway 39

History and Its Significance in the Context of Nagaland and Manipur

Kezhangulie Metha and *Paul B. Chonzik*

Introduction

Road communication system is a primary need for the progress of any region in the world. This is more so in regions where accessibility by train, air and water is not feasible. The hilly regions of North-east India remained isolated from the rest of India and the world for a longer period of time mainly due to this factor. The states of Nagaland and Manipur, for instance, could be opened-up for development only after the Dimapur-Kohima-Imphal cart road came into existence in the later part of the 19th century. Presently, this stretch of road in Nagaland has been numbered National Highway (NH) 29 and the rest from Mao to Imphal and from Imphal to Moreh, NH-2 and NH-102 respectively.The present NH-129 connecting Numaligarh to Dimapur was also a part of the NH-39. The erstwhile (undivided) NH-39 which originated from Numaligarh in Assam to Moreh in Manipur had a total span of 436 kilometres. These highways are now collectively a part of the Asian Highway 1 and 2. The Highway, with connection to the North East Frontier Railway head at Dimapur acts as the primary lifeline for the States of Nagaland and Manipur. It cuts across some important towns and cities like Golaghat, Bokajan, Dimapur, Kohima, Senapati, Kangpokpi, Imphal, Thoubal, Tengnoupal and finally connects India with Myanmar at Moreh. The jungle track of this Highway came into existence following the British expedition against the Nagas after the Anglo-Burmese War of 1824-26. This chapter examines the history and historical significance of this colonial route.

The Historical Background

The process of the British advance into the Hills was slow. It started with a series of explorations for finding out a regular communication with Manipur through the hills. But it was also motivated by a policy to counteract the aggression of the Nagas into the dominion of the East India Company. In January 1832, Captains Jenkins and Pemberton led 700 Manipuri troops with 800 coolies from the Manipur valley, *via* Popolongmai, *Samoogooing* and the Dhunsiri to Mohung Dijooa on the Jumoona (Mackenzie, 1981:101). In between 1832-33, Raja Gumbheer Singh of Manipur, accompanied by the Manipur Levy under Lieutenant Gordon, once again marched through to Assam by a route a few miles to the east of Jenkins' track. Both the parties met with fierce resistance by the Nagas and they had to fight almost every step of the road (*ibid.*,101). In January 1839, Mr. Grange, sub-assistant to the commissioner at Nowgong, was deputed to lead the first expedition into the Angami country (Allen, 1905:11). Thence, ten expeditions were made against the Nagas by the British up to 1850, but without much success. Consequently, in 1851 Lord Dalhousie introduced the Non-interference Policy and further expedition to the Naga Hills were kept in abeyance. Therefore, even though in 1830 a road construction with Manipur was devised it was not until 1870 that it materialised four years after the Naga Hills district was constituted (Bareh, 1970:25).

In 1866, the Naga Hills district was formed with Chumukedima as the Headquarters with Lieutenant Gregory as the officer-in-charge. A road was opened from Dimapur to the new station. The occupation of Chumukedima was followed by the opening of a school and dispensary, the extension of trade and construction of roads to the plains (Mackenzie, *op. cit.*,120-121). The headquarters of the Naga Hills district was later shifted to Wokha but in March 1878 the Chief Commissioner after a personal exploration found Kohimato be the best site for the headquarters of the political officer for commanding the principal Angami and the Manipur frontier line. Kohima was occupied without opposition on 14th November, 1878 (*ibid.*, 132) and with the introduction of effective administration in 1881, hostilities in the Naga Hills gradually disappeared and the Nagas settled down to a peaceful life of cultivation and trade (Milada, 1984:14).

The advent of British administration was followed by the development of bridle tracks across the hills which enabled the government to control the Nagas effectively. A cart road along the jungle routes taken by the early explorers and the expedition parties came into existence. It has been recorded in 1874 that work on the new road up from Dimapur was on (Shakespear, 1883: 146). Major-Gen. Johnstone developed the Diphoo Panee Gorge for the entrance to Nichuguard during the Naga Hills Campaign of 1879-80, and since then it came to be used regularly (Johnstone, 2002: 33). B.C. Allen mentioned that:

> "The only cart road in the district is the section of the great road which runs from Neghereting on the Brahmaputra to Imphal the capital of Manipur. The opening of the Assam-Bengal Railway has rendered Dimapur to all intents and purposes the terminus of the road, and this is the point at which it enters the district of the Naga Hills. The first eight half miles run through the Nambar forest to the foothills at Nichuguard. Here it enters the beautiful gorge of the Diphupani, and the next bungalow is situated at Ghaspani (1,527 feet) 10 miles away. From Ghaspani to Piphima (3,018 feet) nine miles, and from Zubza to Kohima (4,784 feet) ten half miles. Eleven miles beyond Kohima there is a bungalow at Jakhama (5,504 feet), and the frontier district is reached at Mao Thana, nine miles further on. Four miles beyond Mao the summit of the pass is crossed at an elevation of 5,600 feet above the level of the sea, and from this point the road gradually descends to the fertile valley of Manipur. At each of the halting places mentioned there is a comfortable and well-appointed staging bungalow in charge of a *chaukidar*. At present only 14 miles of the road between Mao Thana and Dimapur are metalled, and the un-metalled portions become almost impassable for wheeled traffic during rainy weather" (Allen, *op. cit.*, 56-57).

In the later part of 1901, Lord Curzon, the first Viceroy of India took this route (also known as military road) over the Naga Hills through Kohima to Manipur and on into Burma. His entourage consisted of 49 officials including the Foreign Secretary, Military Secretary, a valet and 38 Indian servants. He returned *via* Chandpur (Bangladesh) and to Silchar by a special train (*The Baptist Missionary Magazine*, 1902: 49; *The Sangai Express*, 29/08/16).

The Deputy Commissioner of Assam in his report of 1901-02 stated that the material prosperity of the Angamis had greatly increased since the advent of the British in the district. By 1903-1904, seventy-three miles of cart-road and about 473 miles of bridle paths were laid in Kohima district (Yonuo, 1978: 110).The opening of the cart road promoted trading activities in the Naga Hills and Manipur. Most of the business was transacted on cash basis, and there was very little barter. The principal trading communities were from Khonoma, Jotsoma, and Mezoma, and men from these villages went as far as Bombay, Calcutta, and Rangoon. Large profits were made by carrying beads through Manipur to Burma, where they were sold for four or five times the amount that had cost. The trade from Manipur passed along the cart road through the district (Allen, *op. cit.*, 59-60). Referring to the Naga Hills and Manipur, in 1901 the London Times had reported that "since they have been brought within the magic circle of *Pax Britanica* they have exhibited a steady and increasing promise of future prosperity" (*The Baptist Missionary Magazine, op. cit.*, 49).

In November 1880, Major General Sir James Johnstone initiated for the improvement of some of the halting places on the way in Manipur. On the request of Johnstone, Sir Bayley, the Chief Commissioner of Assam deputed Lieutenant R.E. Raban for laying the cart road from Manipur valley to Mao. The line from Sengmai was found to be in bad condition throughout, and an exceedingly difficult one in many places. But with the support from Thangal Major and the Maharajah of Manipur and necessary tools, the party examined the whole length of road and trace cutting work was successfully executed with the help of Manipuri workers (Johnstone, *op. cit.*, 173). With the turn of the century and following the building of the railway to Dimapur (also called Manipur road), most people travelled by rail from Guwahati to Dimapur. From Dimapur, the travellers took the pony track up to the Naga Hills and into Imphal valley. And in the years soon after the First World

War travellers went to Kohima, and even Imphal, by pony trap, with baggage following slowly in squeaking bullock cart; but gradually these ramshackle carts were replaced by lorries, the hire of which was a good deal more expensive. The road was single track all the way, a good dry weather surface without tarmac except for one short stretch that was tarred as an experiment by the Burma Oil Company (Evans and Brett, 1962: 5).

The twelve-foot road dominated every day lives of people in the region. It cut across the plain between forest and paddy fields to Nichuguard gate, where a strict one-way traffic control was in order before winding-up the precipitous mountains and rivers to the town of Kohima, 47 miles from Dimapur, and finally to Mao. Mao was a halting place where up and down traffic changed over at a gate (now called Mao Gate) controlled by the police (*ibid.*, 5). The road stretched into Maram and Kangpokpi, sixty miles south of Kohima, and into the plains of Imphal and finally culminates at India-Burma border. The open bag postal system was introduced along the Dimapur-Manipur road; from Dimapur to Kohima it took 12 hours, and again another 5 hours from Kohima to Mao. All the bridle paths in Kohima were under the charge of the Deputy Commissioner. The executive engineer stationed at Kohima was in-charge of important public works in the Naga Hills and Manipur. The principal work the department looked after was the Dimapur-Manipur cart road (Allen, *op. cit.*, 59-68).

In 1942 when the Japanese overran Burma, Kohima developed into a large and important area on the line of communication for the British 4th Corps on Manipur front.The Dimapur-Imphal road writhed its way around Kohima hills and the British had by 1944 established a Traffic Control Post (presently known as TCP Gate) at the main junction on the Ridge (Phillips, 1966: 81). This cart road became a strategically important highway in 1944 when the Japanese and the Indian National Army (INA) launched their campaign into India. The Kohima Ridge is an enormous physical barrier, especially so to an army intent on moving further west towards Dimapur; consequently, its defence and retention was critical if this route was to be barred to an invader (Lyman, 2011: 72). This road features a great deal in the Battle of Kohima (Edwards, 2009: 21). It remained as the only direct transport route from Dimapur to Imphal during the war time. Aerial support apart, this highway turned into a lifeline for both the Allied forces and the Imperial Japanese Army-INA alliance. This was the highway used by the Japanese and the INA in their attempt to change the political destiny of India. This Highway also can be regarded as the main route through which Japan attempted to expand and protect her Greater East Asia Co-Prosperity Sphere. The National Highway 39 turned into a very busy road during and post-Second World War. According to K.N. Pusha of Viswema village, by 1942-43 road widening and repairing work started and a large number of Naga people were employed for the job (Pusha, 2015). More than 50,000 labourers were sent to improve this main road to Burma, which passed through Kohima to Imphal. The Tea-planters of Assam provided half of the labour force (Keane, 2011: 80). Saneilie Sorhienuo of Kohima village was a school student in 1944 and he earned some money every day by helping out in the road construction work after school. He is an eye witness to the countless British military convoys which came up to Kohima along this road (Sorhienuo, 2016). Saneilie and Vipusa Zhashümo of Phesama had witnessed the war and had seen the plight of the Indian refugees coming from Burma (Zhashümo, 2015). According to Saneilie, countless number of refugees came along this highway in haggard conditions. Saneilie and his school friends often saw dead bodies of the refugees from Burma lying on the way sides on their way to and back from school.Vipusa also had seen many Burmese refugees coming through the Manipur-Kohima road that crosses his village. According to him, they appeared dirty, starved, weak and some limping with bleeding feet. He saw some mothers breast-feeding their young ones as they straggled down along the Highway towards Kohima and Dimapur. Vipusa was also only a young boy then and what he had seen haunted him for a long time and he admitted not wanting to re-collect the sight again. On 18 May, 1942, the Governor of Assam reported that approximately 3000 refugees a day were on their way to Dimapur (Keane, *op. cit.*, 30). The march went on day and night along this road and in all, about 1,50,000 Burmese refugees who arrived in Manipur passed through this way to Dimapur and beyond (Evan and Brett, *op. cit.*, 24-25).

In March 1942, the India Command and the Government of India came-up with the urgency for road building programme to connect Assam with Burma and China. The Indian Tea Association wholeheartedly collaborated in the relief works and military projects. They began operations on road construction at the Indo-

Burma border, a project considered to be of highest political and strategic importance (Tyson, 1992: 24). In the same year, on the Manipur front, the influx of the refugees affected the timely development of the bridle-path between Palel and Tamu into a motorable road. Local labourers making the road all absconded after contact with demoralised refugees and on the outbreak of cholera in their camps. It was through the effort of the No.1 Excavating Company and the tea garden coolies that the path was trimmed, drained and rough-surfaced. Most of the work South of Lokchao towards Tamu was cut by hand. However, the original survey for the motorable road had to be abandoned under pressure of time. The engineers started cutting tracks with bulldozers sometimes following the old bridle-path to finish the work within the allotted time of two months. By the 2nd of April, the Lokchao road was completed with a Bailey bridge and with great efforts the road to Tamu was put through on May 5th, a day ahead of the schedule (Evan and Brett, *op. cit.*, 24-28).

The Dimapur-Kohima-Imphal Highway served as one of the main links for the distribution of war relief and had also catered to the refugee camps. Nucleus of the Burma Refugee Organisation (BRC) attended to refugees at Bokajan and Dimapur, located along this highway where thousands were fed, including those who had been working on the road in labour units. Between April and June 1944, the Indian Tea Association and the Burma Refugee Organisation assisted the Deputy Commissioner of the Naga Hills, C.R. Pawsey, with the refugee problem. From 1942 onwards, the main base for the BRC was located near Manipur Road Station [ASA(C) 1944]. It may be noted that the refugee camps were located close to this highway, therefore, this road must have facilitated the saving of thousands of lives.

In the Naga Hills and Manipur, the relief measures post, Second World War were brought *via* this road from the depots at Bokajan. Arthur Swinson recorded that "80,000 coolies were employed in the depots which lined the main roads" (Swinson, 1968: 74). The Fortnightly Report on Assam Relief Measures for the year ending 30/11/44 indicates a daily lift of 30 tons to Kohima and 45 to Imphal up to the middle of November 1944 by this road. The refugees were transported mainly by the 5 and 3 tonners on this Highway. Apart from Dodge, Ford trucks, vehicles provided by the War Transport Department to the Assam Relief Measures, were used between Bokajan, Kohima and Imphal. Lorries, 23 Chassis and trucks were brought from as far as Bombay for relief operation on this highway. Official records also indicates the repatriation of 2174 refugees to Burma from Imphal, Lekhapani, Dimapur and Guwahati between 1945 and 1946 [ASA (C) 1944]. Taking into consideration the number of vehicles used especially by the army and for relief measures, the volume of vehicular traffic cannot be estimated with accuracy. All throughout the war period, the Allied forces attempted to keep open the Dimapur-Kohima-Imphal road (Slim, 1961: 267) as this route remained strategically vital for their success, both at Kohima and Imphal.

Historical Significance of the Highway

In March 1944, when Lieutenant-General Mutaguchi, Commander of the 15th Imperial Japanese Army, launched his three divisions—the 15th, 31st and the 33rd along with Bose's INA into North-east India, this Highway became one of their major targets. While the 33rd and the 15th divisions under the commands of Lieutenant-General Yanagida and Lieutenant-General Yamauchi and a section of the INA were assigned to take on Lieutenant-General Geoffrey Scoones' 4 Corps in Manipur, Lieutenant-General Sato's 31st Division was assigned Kohima and Dimapur. In the assessment of William Slim, Commander of the British 14th Army, Mutaguchi's evaluation of the British position in North-east India revealed that the three key strategic targets in Assam were Imphal, Kohima and Dimapur, the latter of which held stores sufficient to sustain an army on the offensive for several months. In Slim's assessment, if the Japanese captured Kohima, Imphal would be disconnected from the rest of India by land. From the outset Mutaguchi had a plan of taking not only Kohima but Dimapur also. He reasoned that capturing the massive depot in Dimapur would be a devastating, possibly terminal blow to the British ability to defend Imphal, supply Stilwell and mount an offensive into Burma. Mutaguchi also reasoned that with Dimapur captured, Bose and his INA could pour into Bengal, initiating the long-awaited anti-British uprising (Lyman, 2010: 23).

To achieve their primary targets in Manipur, the Japanese sought to take control of the only significant supply road into Imphal (Edwards, *op. cit.*, 11). Therefore, one of the main routes used by the

Japanese and the INA during their entry into Manipur was the Moreh-Imphal stretch of the NH-39 (now NH-102) that connects India with Burma. A section of the Japanese 33rd Division under Major-General Yamamoto fought bitter battles with the British along this highway at Moreh, Lokchao, Shenam, Tengnoupal and between Palel and Imphal. Infantry and tank battles took place along this route, like at Bishenpur, which resulted in heavy casualties and deaths. The Japanese also targeted the northern side of the Imphal-Kohima road (now NH-2 and NH-39) and part of the 15th Division attacked and blocked the point called 'Mission' at Milestone 105. Therefore, the Japanese and the INA terminated the Indo-Burma Road near Kangpokpi on 29th March and from that date until the 23rd June, the plains of Imphal remained disconnected from the rest of India and from its main lifeline from Kohima side (NSA, 1944). Supply to Imphal was maintained by air lifts till the road was re-opened (Lyman, *op. cit.*, 2011: 123).The main objective of the 31st Division was to attack Kohima and prevent reinforcements reaching Imphal (Edwards, *op. cit.*, 53). During the first week of April, the Japanese soldiers attempted to block the NH-39 between Kohima and Dimapur at milestone 36, just to the north of Zubza to stop reinforcement reaching Kohima (Swinson, *op. cit.*, 95). At Kohima, some of the fiercest battles of the Second World War were fought alongside the length of this Highway. A good many military installations, official buildings, hospitals, supply depots were located not very far from this Highway. The battle of the Tennis Court, which sealed the fate of the Japanese and the INA at Kohima, was fought above this highway. The Japanese army's repeated attempts to terminate the Dimapur-Kohima road and to dislodge the British position at Zubza where the NH-39 passes through towards Dimapur, met with failure.

The twin battle of Imphal and Kohima in 1944 has been adjudged as the "greatest ever battle involving British forces" (PTI-NP, 2013). Mountbatten, the Supreme Commander of South-East Asia also has described the battle of Kohima as "probably one of the greatest battles in history" (Edwards, *op. cit.*, 11). And in the words of the Field Marshall William Slim, the Imphal-Kohima battle was the "last and greatest of the series on all the Burma fronts" (Slim, *op. cit.*, 307). The battle of Imphal-Kohima had proven otherwise the myth of "Japanese Invincibility" for the first time, and with it ended the INA march on to Delhi.

Conclusion

It may be stated that the development of the Dimapur-Kohima-Imphal cart road and the bridle paths in the hills strengthened the British administration in the Naga Hills and Manipur. It led to further annexation of tribal areas, ushering in peace and progress. The opening of roads led to growth of trading activities and consequently promoted money economy in the Naga Hills and Manipur. More money came into the hands of the people during the construction of this road which also promoted business and other economic activities in the subsequent years. This road contributed immensely towards the victory of the Allied forces at Kohima and Imphal in 1944. And considering the geostrategic location of North-east India, this highway is facilitating the safety of India's Eastern borders. At present, this Highway is maintained by the Border Road Task Force and it continues to play important roles in the overall aspects of life of the people. It has been promoting socio-cultural relationship between the people of the two states and beyond for over a century now. Above all, the Highway is contributing immensely towards the security and economic development in North-east region, especially in Nagaland and Manipur.

Bibliography

1. Archival Sources

Assam State Archive (Confidential) Governor's Secretariat, 1944, *F/No. 122-c/44.*

Assam State Archive, Secretariat Administration (Record & Library) Department, Assam Secretariat (c), *File No. 5266.*

Government of Nagaland State Archive, Kohima, The Japanese Campaign of 1944 (Confidential File No. 497), 1944.

2. Oral Sources

Pusha, K.N., Viswema village, Kohima district, Nagaland, 2015.

Sorhienuo, Saneilie, Kohima village, Kohima district, Nagaland, 2016.

Zhashümo, Vipusa, Phesama village, Kohima district, Nagaland, 2015.

3. News and Magazines

London, APR 21 (PTI): Second World War Kohima Battle is Britain's Greatest Battle (Published in *Nagaland Post*, April 22, 2013, Vol. XXIII, No. 135).

The Baptist Missionary Magazine (1902), Vol. LXXXII, published by the American Baptist Missionary Union, Boston.

4. Internet Resources

The Sangai Express, Manipur, www.thesangaiexpress.com (29/08/16, accessed at 12:35 am).

5. Government Publications

Bareh, Hamlet (1970): *Gazetteer of India*, Nagaland, Kohima District (Government of Nagaland, Kohima).

Verrier, Elwin (1961): *Nagaland* (Published by P. Dutta for Research Department, Adviser's Secretariat, Shillong).

6. Books

Allen, B.C., (1905): *The Naga Hills and Manipur: Socio-Economic History* (Name of publisher/s not mentioned, Shillong).

Edwards, Leslie (2009): *Kohima* (The History Press, UK).

Evans, Geoffrey and Brett-James (1962): *Imphal* (MacMillan & Co., London).

Ganguli, Milada (1984): *A Pilgrimage to the Nagas* (Oxford & IBH Publishing Co., New Delhi).

Keane, Fergal (2011): *Road of Bones* (Harper Press, London).

Lyman, Robert (2010): *Kohima 1944* (Osprey Publishing, GB).

—(2011): *Japan's Last Bid for Victory* (The Praetorian Press, GB).

Johnstone, James (2002): *Manipur and the Naga Hills* (Gyan Publishing House, New Delhi).

Mackenzie, Alexander (1981): *The North-East Frontier of India* (Mittal Publications, Delhi).

Phillips, C.E. Lucas (1966): *Springboard to Victory* (William Heinermann, London).

Shakespear, L.W. (1983): *History of the Assam Rifles* (Cultural Publishing House, Delhi).

Slim, William (1961): *Defeat into Victory* (Cassell & Company, GB).

Swinson, Arthur (1968): *Kohima* (Arrow Books, London).

Tyson, Geoffrey (1992): *Forgotten Frontier* (Spectrum Publications, Guwahati).

Yonuo, Asoso (1978): *The Rising Nagas* (Manas Publication, Delhi).

7. Other References

Barker, A.J. (1963): *The March on Delhi* (Faber and Faber, London).

Bond, B. and Tachikawa, K. (2004): *British and Japanese Military Leadership in the Far Eastern War, 1941-1945* (Routledge, New York).

Campbell, Arthur (1957): *The Siege* (George Allen & Unwin Ltd., GB).

Chakravarty, Sudeep (2012): *Highway 39* (Harper Collins, India).

Chapman, Joyce Lebra (1975): *Japan's Greater East Asia Co-Prosperity Sphere in World War II* (OUP, Kuala Lumpur).

McLynn, Frank (2011): *The Burma Campaign* (Yale University Press, USA).

Reid, Robert (1983): *History of the Frontier Areas Bordering on Assam, 1883–1941* (Eastern Publishing House, Delhi).

Verrier, Elwin (ed.) 1969: *The Nagas in the Nineteenth Century* (OUP, London).

Webster, Donavan (2004): *The Burma Road* (Harper Perennial, USA).

17

The Importance of Roads in the Spread of Christianity and Western Education

A Review on the Naga Hills of North-East India

Rosanna Kikon

Ever since ancient times, roads have played a very important role in the advance of human communities. One of the main features of roads is the promotion of trade and communications between peoples of different cultures thereby influencing its social, economic and political development. Another important feature is its role in bringing peoples of different cultures and religions to intermingle, leading to the diffusion of religious and cultural ideas. In the light of this, the Chapter explores how roads helped the missionaries find their way into the Naga Hills and how they transmitted Christian beliefs and practices, accompanied by education, amongst the head-hunting tribes of the Naga Hills, now known as Nagaland, in North East India in the later part of the 19th Century.

The first forms of roads in the Naga Hills, interconnecting the villages, were mere jungle trails, called Naga paths, which were taken-up one side of the hill and down the other. These tracks were not planned and constructed by road engineers, but came into being due to regular use over time to get from one place to another, and were confined to walking, for humans carrying loads on backs and heads. An early description of Naga paths can be found in the government survey records and missionary journals. One description as given by Woodthorpe reads:

> "The path ran along the top of a very narrow ridge, following all its natural features...one false step would have sent us over the precipices into the unknown depths of jungle below...to crown all, the path was thickly studded with 'panjis' for two-thirds of the way...outside the village, on either side of the path were very nasty pits about 5 feet deep and 34 feet square, and bristling with 'panjis' 3 feet long" (Thuillier, 1876:57).

The social practices of the time made it necessary for the Nagas to adopt such rigorous measures for safety, but extensive regions lay beyond their control, and often they fell victims to headhunting during their trips. As dangerous as it may have been, these jungle tracks were the only routes through which the tribal villages kept in touch; serving also as trade routes connecting them to Assam. Early records state that since the very early times the Nagas were used to visit the markets of Jorhat, Kacharihat and Golaghat and cotton formed the chief articles of their commerce (Barpujari, 2003:219) and the extensive trade in salt of the Konyak Nagas (BMM, 1839:148). Even till the time of the British expedition of 1880, the paths were few while the cart-road was non-existent except for the cart-road which ran from Dimapur to Manipur *via* Kohima. There was a track from Samaguting to Piphimá, Nerhema and Wokha, referred to as 'political path' which not only climbed steep slopes and descended into deep ravines, but were so closely shut in with dense jungle that troops were compelled to move along them in single file (Intelligence Branch, 1907:201).

The American missionaries made a three-pronged entry into the Naga Hills with their evangelism. In 1875, Reverend E.W. Clark entered the Ao country, which was then beyond the pale of British control, taking the route from Sibsagar-Amguri in Assam (Phillip, 1999:56) to the Ao country and set-up mission at Deka Haimong. Mission entry in the Angami and Lotha country was made possible through the government opened roads—the Reverend C.D. King entered the Angami country through Dimapur-Chumukedima-Kohima road and set up mission at Kohima in 1880; the Wokha mission was started by Reverend W.E. Witter, having entered through the Golaghat-Merapani-Wokha road in 1885. Early missions encompassed a centre and out-stations and work for the missionaries was not easy in the early stages because of difficulties of communications. The journey into the Naga Hills was not easy; jungle trails were infested with wild beasts and swollen rivers, over which there were no bridges, had to be crossed. Sometimes they had to make their way up trails that were mere ledges in the rocks, clinging to roots and grasses as they went (Vickland, 1926:57-58). The missionary, Clark, in his 1881 report of the Ao country stated that, "the escort was soon struck with fear by the evidences of ambuscading along the route and some of them said they would be thankful if they got back home with their lives" (BMM, 1881:320). Tribal wars that existed between communities were a security risk and, coupled with the threat of wild animals, hindered their travel and the widespread expansion of the Gospel.

A positive factor for the missionary operations was the extension of the British rule in the region. The British secured initial control of Chumukedima as the Naga Hills headquarters in 1866 (Administration Report, 1896:92) ultimately leading to the formation of the Naga Hills district with Kohima and Wokha in 1878 (Reid, 1983:101). In 1890, Mokokchung, having been formally brought under British administered area in 1889, the headquarter at Wokha was shifted to Mokokchung to serve the dual purpose of administering Wokha, as well as the frontier tribes. In 1903, the eastern Angami villages and the areas consisting of the Semas was brought under political control (*ibid.*,125-133).The British political control over the Lotha tribe under a Sub-Divisional Officer with a native military force made Wokha a comparatively safe place for mission work. The bridle path running from Golaghat in Assam to Wokha rendered the Wokha mission station more desirable as the connecting link between the Angamis and Aos. With Kohima sixty miles south-west, a grand trunk line of missions to the Nagas could be opened (Clark, 1907:150). The Wokha station was also to serve as the base for outreach to the Sema tribe by the missionaries. The establishment of missions at Kohima (1880) and Ukhrul in Manipur (1896) both of which were on or near the high road between Assam and Burma became a point of contact of the Baptist mission work in Assam and Burma (Merriam, 1913:128). The Kohima field was centrally located and work was carried out for six tribes—Angami, Chakrumia, Rengma, Kuki, Kacha Nagas, Sema and some Assamese (ABFMS, 1929:107).

After the preliminary work of subjugating the Nagas, the next important task before the British was to secure communication by means of military roads between carefully selected strategic points in the administered areas. The development of the bridle paths enabled the extension of British political influence across the Naga Hills, and the British replaced their military control by a policy of indirect rule through administration. Road building was important for the functioning of an efficient administration—to effectively govern, communicate and control. Village roads were constructed and communications in the hills and between the hills and the plains were opened. The road from Golaghat to Wokha which ran about 62 miles in length was made in 1876 (Shakespear, 1983:145). Between 1893-94 roads from Mokokchung to Mariani, Mokokchung to Tamlu and Mokokchung to Doyang were opened and several existing roads were widened. The iron suspension bridge over the Doyang river, linking the Lotha area to the Ao area through the Sema area, was completed (NSA General Dept.). In 1903-04, seventy three miles of cart-road and about 473 miles of bridle paths were opened and maintained in the district. The cart-road from Manipur to Dimapur, where the Assam general railway connected, through Kohima, was widened (Yonuo, 1974:110). All the outposts were connected with the headquarter station at Kohima by good, well-graded bridle paths which extended in several directions into the Sema hills with comfortable rest-houses at all stages (Shakespear, 1914:227). By 1920 there were 740.83 miles of bridle paths and construction of new roads and bridges were undertaken (NSA Political Branch). Large areas could now be very easily governed with the help of good transportation system. These expensive projects were undertaken primarily, for administrative exigency, which in turn, encouraged travel and commerce. Consequently, roads that served the British interests and motives also helped facilitate the spread of cultural and religious ideas into the Naga Hills.

Table 17.1 shows the spread of Christianity in the Naga Hills after the road development works undertaken by the British Government in the administered areas.

Once new Naga areas were brought under British control roads became safer to travel resulting in the growth of Christian population. When the Ao country was formally annexed in 1889 mission work could then be carried out anywhere in the tribe with safety (BMM, 1889:259-260) and Christianity started entering new

Table 17.1: Growth of Christian Population: Spread of Christianity in Naga Hills—1891–1931

Year	*Total Population*	*Christian Population*
1891	1,22,867	231
1901	1,02,402	601
1911	1,49,623	3,308
1921	1,60,960	8,734
1931	1,78,844	22,908
1941	1,89,641	34,000

Source: *Census of 1891,* Vol.I, *Assam Report:* 57, 88; 1901,Vol. IV, Part I, *Assam Report:* 32, 52; 1911, Vol. III, Part I, *Assam Report:* 24, 38; 1921, Vol. III, Part I, *Assam Report:* 26, 61; *Census of 1941:* 8 & 1931, Vol. III, Part I, *Assam Report* 200; *Census of 1941,* Vol. IX, Assam: 2, 23.

villages. The Census of 1901 projects the growth in the number of Christians to almost triple the number of 1891. The mission centre was shifted from Molung to Impur in 1893, which was more centrally located and near the government roads (ABMU, 1894:101) which connected Sibsagor, Impur, Wokha, Kohima and Sylhet. There was also the advantage of a cross road to the Assam Valley, over which all supplies could be obtained (BMM, 1895:307). The Impur station catered to the Ao, Lotha, Sema, Sangtam, Chang and Phom tribes (Downs, 1983:139). The quantum increase in the Christian population in 1911 resulted due to the inclusion of the Eastern Angamis and Semas which was added to the Naga Hills since 1901 (Census, 1911:1). According to the Census of 1921, the Naga Hills increased by 9,910, Kohima subdivision gaining 2,894 and Mokokchung 7,016. About 7,800 trans-Dikhu Konyaks and others were included for the first time and censused in Mokokchung Sub-division (Census, 1921:19). The increase in the Christian population as shown in the 1931 Census can be attributed to the development of road connectivity in the Sema country. The missionary report of 1926 corroborates the growth of Christianity among the Sema Nagas (ABMU Jorhat Conference, 1926:3) who lived mainly between the Kohima and Impur fields. The opening of the suspension bridge over the Doyang river made possible the flow of human traffic all throughout the year, when earlier it remained inaccessible during the monsoon season. Christianity thus spread across the hills as colonial developments increased mobility.

The establishment of the posts of colonial administration facilitated and protected the missionary activities in the administered area. Among the main criteria for the choice of the permanent locations for the mission stations were the favourable conditions for the health and safety of the missionaries and access to supplies. In the subsequent decades, these mission stations became the centres of religious, economic, medical and educational life in the areas. Mission work for the Naga Hills was supervised from Kohima and Impur, Wokha station having been shortlived. These stations were staffed with mission workers who carried out tours of the surrounding areas. However, away from the government roads, most villages were still accessible only by minor jungle paths. Even in such circumstances, the jungle tracks which had at one point of time brought many a headhunting Naga into combat with one another, were now treaded upon for a peaceful purpose. Many walked for long miles into Kohima and Impur to be baptized as Christians. Gradually, native converts who studied at Impur began to disseminate the Gospel in their native villages and beyond. Missionary records mention of how an Ao convert, Intionen, after a course in Impur school, returned to his own village as a teacher and founded a church. In time he left the charge of his village church to the converts and he ventured forth to carry on pioneer work in a new and distant village. Converts from other tribes, such as the Lothas, studying at Impur school, also took the Gospel to their villages during vacations and returned to Impur with men and boys for Baptism (BMM, 1904:106-07). From these stations numerous tribes had been touched, so that, the Nagas "once held apart by linguistic differences and head hunting proclivities were beginning to feel the bond of Christian brotherhood" (Missions, 1921:399). Baptist associations, which were held annually, became a huge affair for it saw the congregation of multitudes of Naga Christians—men, women and children, who walked miles from far-flung villages to worship in unity, a union made possible due to the existence of safe roads and the spread of Christianity. According to the missionary reports, by 1944, though travel among the border tribes like Sangtams, Konyaks, Changs and Phoms were restricted by Government work among them was carried out by the native Ao Christians (ABMC Report, 1944:20).

In order to spread the Gospel, the missionaries gave importance to teaching the natives reading and writing. Education was initially left in the hands of the missionaries with aids from the Government. The spread of Christianity and education was made possible by the making of government roads, opening the way for access into the distant villages. This encouraged the missionaries to open schools in different villages for the purpose of spreading Christianity and education. Village schools formed the chief institution for education and conversion. In 1898, a training school for the native Nagas was opened at Impur and this remained the only school for furnishing teachers to the several Naga tribes; to train native teachers to spread the Gospel (BMM, 1905:19). Impur became the hub of education where Naga tribes from different parts of the Hills convoked to learn. Stress was laid on writing, which, accomplished with great difficulty, soon awakened an interest in education throughout the hills. As the missionary S.A. Perrine wrote, it led "to a revival in 'letters'—a Naga 'renaissance' so to speak. So many and urgent became the letters flying back and forth on matters that were hardly urgent, that we had to interfere in the interest of the village letter carriers. A Naga dak is now talked of" (ABMU, 1899:40). The villages were miles apart and the only route was the jungle track which descended hundreds and thousands of feet into the valley and up again as far. Despite this hindrance, the Impur training school had representatives from six tribes. Boys travelled in groups to attend the mission school at Impur, for though road connectivity improved there was always the fear of attacks by wild animals. As the missionary, Dowd reported, more than two fifths of the pupils were from foreign tribes. The majority of which were Lothas and a good number of Semas represented in smaller number by other tribes. The boys quickly learned the Ao dialect and thus had access to a part of the Scriptures. Some of the foreign pupils were found to be in no way behind the Ao pupils who studied in their own language (BMM, 1904:106). Therefore, one area of culture influence would be seen in language. The Ao dialect, being the first to be written in the Roman script, became the dominant dialect of the Impur School. The same was the case with students from other tribes studying in the Kohima mission school; students had to learn the Angami dialect to study the scriptures. It was a proof that not only was religion and education promoted, but it resulted in the growth of frequent contact between the erstwhile 'isolated' Naga tribes and the beginning of cultural diffusion. Roads connecting the Hills to

Assam also opened the way for students to study in the mission schools in Guwahati, Jorhat, and Shillong (ABM Jorhat Conference, 1926:65), leading to a convergence of culture. The association of the Nagas, with their neighbour, Assam had a very lasting impression, and that is, the birth of Nagamese, the common pidgin, which is spoken amongst the Nagas today.

Since the beginning of the mission in the Naga Hills, the missionaries had traversed the Naga roads mostly on foot, and on pony or horse backs to reach the people with religion and education. But over the decades changes began to take place and in 1930, an historic event took place when the missionary, G.W. Supplee, conquered the Naga roads on a motorcycle when he made a mission trip *via* Nakachari on the plains to Impur and back to Kohima *via* Wokha (ABFMS, 1930:119). By the time, the last of the missionaries left the Naga Hills in the early 1950s, motor vehicles had started plying on these roads.

It can be concluded that, roads which had at one point of time served the British colonial administration and the missionaries in their evangelization, opened a vista of progress for the Nagas. By the turn of the century, sizable Christian communities flourished throughout the Naga Hills and education began to spread in earnest. Over the century, bridle paths have developed into motorable roads, linking every village, all throughout the state. Presently, the road from Manipur to Dimapur through Kohima and the Kohima-Mokokchung road *via* Wokha now fall under National Highway No. 29 and 2 respectively, and serve as the primary life-lines of the state. The Golaghat-Merapani-Wokha road not only serves as a trading route, but also promotes cultural meets between Naga tribes and the people of neighbouring Assam, to foster friendship and solidarity. These roads, thus, not only cater to trade and commerce but also serve as an agent of cultural diffusion within Nagaland and beyond.

Bibliography

Archival Sources

General Administration Report of the Mokokchung Sub-Division 1893-94 (General Department, File No. 190) Kohima: Nagaland State Archives (NSA) viewed on 7 August, 2015.

Annual Administration Report of the Naga Hills District for 1920-21 (Political Branch: File No. 395) Kohima: Nagaland State Archives (NSA) viewed on 7 August, 2015.

Government Publications

Government of Nagaland (2013): *Human Development Report Wokha District* (Kohima).

Census of India (1892): Vol. I, Assam Report 1891 (Shillong: Assam Secretariat Printing Office).

Census of India (1902): Vol. IV, Pt. I, Assam Report 1901 (Shillong: Assam Secretariat Printing Office).

Census of India (1912): Vol. III, Pt. I, Assam Report 1911 (Shillong: Assam Secretariat Printing Office).

Census of India (1923): Vol. III, Pt. I, Assam Report 1921 (Shillong: Government Press).

Census of India (1932): Vol. III, Pt. I, Assam Report 1931 (Shillong: Assam Government Press).

Census of India (1942): Vol. IX, Assam 1941 (Delhi: Manager of Publications).

Thuillier, Col. H.L. (1876): *General Report on the Topographical Surveys of India, 1874-75* (Calcutta: Office of Superintendent of Government Printing).

Reprint of Report on the Administration of the Province of Assam 1892-93 (1896): *Physical and Political Geography of Assam* (Shillong: Assam Secretariat Printing Office).

Intelligence Branch, Division of the Chief of the Staff, Army Headquarters, India (1907): *Frontier and Overseas Expeditions from India*, Vol. IV, North & North Eastern Frontier Tribes (Simla: Government Monotype Press).

Books

Barpujari, S.K. (2003): *The Nagas: The Evolution of their History and Administration* (Guwahati: Spectrum Publications).

Clark, Mary Mead (1907): *A Corner in India* (Philadelphia: American Baptist Publication Society).

Downs, F.S. (1983): *Christianity in North-East India* (Guwahati : Indian Society for Promoting Christian Knowledge).

Merriam, Edmund F. (1913): *A History of American Baptist Missions* (Philadelphia: American Baptist Publications Society).

Phillip, P.T. (1999): *The Growth of Baptist Churches in Nagaland* (Guwahati: Christian Literature Centre).

Reid, Robert (1983): *History of the Frontier Areas Bordering on Assam* (Delhi: Eastern Publishing House).

Shakespear, L.W. (1914): *History of Upper Assam, Upper Burma and North Eastern Frontier* (London: MacMillan).
Shakespear, L.W. (1983): *History of Assam Rifles* (Delhi: Cultural Publishing House).
Vickland, E. (1926): *With Christ in Assam* (Philadelphia: The Judson Press).
Yonuo, Asoso (1974): *The Rising Nagas* (Delhi: Vivek Publishing House).

Magazines

American Baptist Missionary Union (1894): *80th Annual Report* (Boston: Missionary Rooms).
American Baptist Work in Assam (1926): *Jorhat Conference 1926* (Jorhat: The Assam Printing Works).
American Baptist Foreign Mission Society (1929): *115th Annual Report* (New York: Foreign Mission Quarters).
American Baptist Foreign Mission Society (1930): *116th Annual Report* (New York: Foreign Mission Headquarters).
American Baptist Mission Society (1944): *Assam Baptist Missionary Conference Report*, 41st Session, Nowgong, Assam, (Calcutta: Baptist Mission Press).
Baptist Missionary Magazine (1839): Vol. XIX, No. 6, June, 1839 (Boston: American Baptist Missionary Union, Boston).
Baptist Missionary Magazine (1881): Vol. LXI, No. 10, September (Boston: American Baptist Missionary Union).
Baptist Missionary Magazine (1889): Vol. LXIX, No. 7 (Boston: American Baptist Missionary Union).
Baptist Missionary Magazine (1895): Vol. LXXV, No. 7 (Boston: American Baptist Missionary Union).
Baptist Missionary Magazine (1904): 90th Annual Report, Vol. LXXXIV, July, 1904 (Boston: Missionary Rooms).
Baptist Missionary Magazine (1905): Vol. LXXXVI, No. 1 (Boston: American Baptist Missionary Union).
Missions (1921): Vol. 12, No. 7 (New York: General Board of Promotion of the Northern Baptist Convention).
The Assam Mission of the American Baptist Missionary Union (1899): 5th Triennial Conference 1899 (Calcutta: Baptist Mission Press).

18

Effect of Road Transport in the Development of Rural Economy of North-East India

Subit Dutta

Introduction

The North-east region of India comprises Arunachal Pradesh, Assam, Manipur, Meghalaya, Mizoram, Nagaland, Tripura and Sikkim. This region has abundant natural resources like—minerals, coal, petroleum, forest, water, and biodiversity which constitute the basis of its potential wealth. It also provides a platform for India to extend its linkages with the immediate neighbourhood such as—China, Bhutan, Bangladesh, and Myanmar and thus influence the South-east Asia. But the region is also wrecked by several problems, the most invasive being underdevelopment, a huge number of insurgent outfits, and the resultant vulnerability to external interference. Defence and development of the North-east have thus emerged as a fundamental aspect of national security. The development of an efficient road network in the region is the only way to overcome these problems. Roads not only help to bind the divergent regions but also open up new areas for settlement and economic ventures.

The pattern of the road network that has evolved in the North-east reflects the priorities that have driven road building in the last 70 years. In the immediate aftermath of independence, there was a felt need to establish severed linkages between the North-east and the rest of the country. The eruption of the border dispute with China drove the Government of India to build roads along the borders. Subsequently, the emergence of new political entities in the North-east, multiple insurgencies, and pervasive under-development led to a change in focus towards developing intra-region connectivity. Today, the rapid pace of the Chinese road building along and across its borders, especially in Myanmar, has led to a renewed focus on building strategic and trans-border roads. Another factor that has contributed to the focus upon building trans-border roads as well as an overall road network in the North-east is the realization that the region can serve as a bridge between the Indian economy and the vibrant economies of the East and South-east Asia. If the above-mentioned factors impelled the Government of India to build a road network in the North-east, there were many hurdles which acted as brakes to its development. Some of these impediments were inadequate funds, vested interests, faulty policies, insurgency etc. which halted the smooth development of an efficient road network in the North-east.

Objectives of the Study

1. To examine the factors influencing the development of road network in North-east India.
2. To examine the challenges to the development of the road network in North-east India.

Methodology

The study is descriptive in nature and based on the secondary data which has been collected from various Annual Reports, Journals and websites.

Factors Influencing the Development of Road Network

Establishment of Severed Links

An important challenge that India in general and North-east India in particular faced after independence, was the reinstatement of brutal communication links. During the colonial period, the British had access to the region through undivided Bengal and, as a result, all transportation lines were routed through that province. Thus, the Garo, Khasi and Lushai Hills, Tripura, and Assam including Cachar were connected with Bengal. As British administration penetrated further into the frontier areas, the Naga Hills and North East Frontier Tracts were also connected by roads that ran through the Brahmaputra Valley. Partition disrupted this colonial pattern of transportation, which not only cut the region off from the rest of India, but also led to increased transportation costs and dislocated traditional markets. Faced with the identical challenges of integrating this region and assuring its economic growth, independent India accorded top priority to the construction of roads in North-east India. The first important step was to connect Assam with the rest of India. For that, the existing National Highway-31 was realigned by laying a road from Bihar to Assam *via* Siliguri, avoiding East Pakistan. The second step involved connecting Tripura and Khasi and Garo Hills with the Assam road system and then with the rest of India.

Various other projects were also initiated to enhance intra-region connectivity in the North-east. It is important to note that these roads also served other objectives. A diversion between Litan and Ukhrul was constructed primarily on NH-39, which connects Dimapur with Imphal, "to combat the communist activity amongst the Ukhrul Nagas". Likewise, the Udaipur-Sabroom, Bagafa-Belonia, and the Khowai-Teliamura roads were constructed "to serve the displaced persons in Tripura" (*Annual Report 1950–51)*. Gradually, other areas of the North-east were also brought within the road network. But majority of the roads built were dirt roads that were easily washed away during heavy rainfall North-east receives the heaviest amount of rainfall in India.

Border Dispute with China

In 1959, China formally laid claim to Arunachal Pradesh, marking the beginning of a festering border dispute. Border clashes with China led to the Indian army taking over the construction of roads in Arunachal Pradesh. While roads and tracks were being built in the border areas, a decision was taken not to construct roads and airfields within 30 miles of the border. The short-sighted argument forwarded was that "connecting forward areas with the interior would serve Chinese offensive purposes more than India's defensive purpose" *(Report of Administrative Reform Commission, 2001)*. It is now well known that this strategy did not stop the Chinese in 1962, when they steam-rolled all the way to the outskirts of Tezpur in Assam.

In the aftermath of the war, the government realized the need to establish better connectivity in Arunachal Pradesh and started building roads in the state. But the roads laid were mostly concentrated along foothills and were primarily aimed at providing lateral movement to the troops. However, the areas bordering China were still left undeveloped and devoid of roads.

Reorganization of States

A new phase of road building started with the reorganization of the North-eastern states in 1972, when seven political units with 30 districts emerged in the region. The newly formed states threw up demands for the development of road networks. They wanted their capitals to be connected with district headquarters as well as with rest of the country. The objective was to connect remote areas with the administrative centres to facilitate the extension of state administration as well as to extend development programmes. Thus, the immediate threefold task in road building were: linking each district headquarters with the state capital, linking each state capital of North-east, and improving the links of Assam with rest of India (Chaube, 1975).

Insurgency

Road building in the region has also been impelled by various insurgent movements that have plagued the region since independence. Since the inhospitable terrain and the lack of a road network in remote areas proved advantageous for the insurgents, the government funded to construct new roads in the insurgent-affected areas. The initiative was based on the understanding that roads are the biggest enemy of insurgents.

However, once insurgents gave-up the gun, road building gathered momentum mainly because one of the principal demands of the former rebels was economic development. For that, various peace accords signed between the Indian Government and former rebels in the North-east contained the seeds for the further spread of the road network. The Assam Accord of 1985 (*Report of Assam Accord of 1985*) focused on the improvement of infrastructure in the state.

Economic Development

The state of pervasive underdevelopment in the North-east is often cited as one of the reasons for the persistence of insurgency in the region. And the Central Government, over the years, has been trying to ameliorate the situation through various developmental programmes with little success. In February 1990, a committee under the chairmanship of L.C. Jain was constituted by the Central Government to study the infrastructural requirements of the region. Another high level commission under the Chairmanship of S.P. Shukla was later set-up in 1997 to tackle the problem of the backlog in basic minimum services and infrastructure needs of the North-east (Planning Commission, Government of India, 1997). The commission noted the inadequate infrastructure in the region and strongly advocated the need to develop them. Based on their recommendations, the Prime Minister's new initiative for the North-east was formulated in 1996.

The North-east has also received generous funding from various schemes and organizations for the development of the road network. Some of them are the Pradhan Mantri Gram Sadak Yojana (PMGSY), the North Eastern Council (NEC), and the Ministry of DONER. The PMGSY is a centrally funded scheme to provide connectivity to rural habitation as part of a poverty reduction strategy. In the North-east, there are many rural road projects which are funded under this scheme (http://pmgsy.nic.in). Various schemes of Central Government of India as well as the states have contributed much to the improvement of the road network in the North-east. However, a look at the density of roads reveals that the development of roads has been uneven in the region.

Chinese Road-building Activities

Along with internal imperatives, there are also external factors which influenced India's decision to build roads in the North-east. China has been building roads to consolidate its hold over border regions and integrate them. This has been especially so in Tibet, which borders India. Today, Tibet has five NHs along with 15 trunk highways and 315 feeder roads. The whole length of highways open to traffic in Tibet is more than 40,000 km (Raja Mohan, 2006). Though China has developed many transport links in its neighbouring countries, those developed in Myanmar and Bangladesh are of importance as far as India's North-east is concerned. China's plan to access the Bay of Bengal *via* the Irrawady Corridor is one such initiative. This involves the construction and upgrading of the 'Burma Road' from Kunming to Mandalay to Ruili and the improvement of the railway line from Dali to Ruili (John, 2006).

Recent Chinese reassertion of claims over Arunachal Pradesh with repeated intrusions by People's Liberation Army (PLA) units have led to the Government of India instructing the Ministry of Roads, Transport and Highways to prepare a revised plan by merging the first and second phases of SARDP-NE. This would result in the construction of an estimated 3,062 km of roads and their completion date is being advanced to 2009. Of these, nearly 40 per cent have been marked as 'strategically important'. In March 2008, the Ministry of Defence directed the Border Roads Organisation (BRO) to concentrate more on building roads in Arunachal Pradesh.

Trans-Border Connectivity

The construction of an efficient road system both in the North-east and across its border has also become necessary to realize the potential of India's goal of economically linking-up with East and South-east Asia through the Look East Policy (LEP). The LEP, in turn, has fostered many other regional cooperative initiatives such as Mekong Ganga Cooperation (MGC), the Bay of Bengal Multi Sectoral Technology and Economic Cooperation (BIMSTEC), the Bangladesh–China–India–Myanmar Regional Economic Forum (BCIM), and the Asia Land Transport and Infrastructure Development (ALTID) programme for the development of the entire region (*Report of the Asian Dialogue Society's Greater Asia Initiative*, 2006). Though the LEP was initiated in 1992, the North-east did not figure as a region through which economic interactions with South-east Asian countries could be routed given its dismal state of road connectivity. For instance, out of a road network of 82,000 km in 2005, only around 25,000 km were paved. The rest were unpaved and thus unsuitable for heavy traffic. As far as NHs are concerned, out of 6,880 km, 3,725 km were less than two-lane. State capitals like Agartala and Aizawl are not even connected by a two-lane NH. Some 62 district headquarters are not connected by a two-lane road. Border areas and other remote and backward areas of the North-east are even worse and connectivity to neighbouring countries has been minimal.

Nevertheless, the fact that the North-east, by virtue of its geographical proximity to the South-east Asian region, provides India with a 'land bridge' to Myanmar and a gateway to South-east Asia cannot be overlooked. It is this realization that led to the Government of India's decision not only to connect the region with the rest of the country with highspeed, high-capacity highways, but also to improve the intra-region and across-border connectivity. It has also reserved 10 per cent of the allocations for road development to the North-east (*Annual Report 2006–07*).

To improve road connectivity in the North-east and also with the rest of the country, the Government of India has launched several important programmes like (1) National Highway Development Programme-Phase II (NHDP-II), (2) National Highway Development Programme-Phase III B (NHDP-III B), (3) Special Accelerated Road Development Programme for the North-east Region (SARDPNE). These programmes are being implemented by different agencies such as the National Highways Authority of India (NHAI), BRO, and State PWDs.

The Central Government also plans to invest nearly Rs. 5,000 crores on projects relating to road connectivity in the region over a span of five years during the Eleventh Five Year Plan. Once completed, these projects are expected to considerably improve connectivity by adding substantial kilometres of roads of different categories. State capitals and district headquarters would be connected with four-lane highways and two-lane roads, respectively. It is expected to provide road connectivity to backward regions and improve roads of strategic importance in border areas. Apart from these projects that aim to improve inter- and intra-region connectivity, the Government of India has also undertaken the construction of some trans-border road projects to establish connectivity with the East and South-east Asian countries. These include: (1) Tamu–Kalewa–Kalemyo Road: This 160 km road is an extension of NH-39 connecting Numaligarh–

Kohima–Imphal. Opened for traffic in 2001, it forms an important link from the India–Myanmar border (Moreh) to central Myanmar and the commercial and cultural centre of Mandalay (*Annual Report 2000–2001*). (2) Rih–Tiddim and Rhi–Falam Road: The BRO has undertaken to construct this 225 km road inside Myanmar from Rih–Tiddim to Rih–Falam. They would improve connectivity between Champai in Mizoram and Rih in Myanmar, which had started functioning as border trade posts from January 2004. India and Myanmar have also agreed to begin the construction of a trans-border road connecting the Chapi village of Mizoram and Sabawngte village and the Matupi Township of Chin state (*Khonumthung News Group*, 2006). (3) Trilateral Highway: India plans to connect the North-east region with Thailand through a 1,360 km highway. This highway would connect Moreh to Maesot in Thailand *via* Bagan in Myanmar. Construction of the highway has already begun. It is reported that eventually it would connect India with Vietnam, Laos, and Singapore (*India Business News*, 2007). (4) Asian Highway: The proposed Asian Highway Network has a total length of 1,41,000 km and covers 32 Asian countries. Of the various routes of this highway, route number A-1 would traverse through North-east India. It would enter the region at Dawki in Meghalaya from Bangladesh and after passing through four states in the region would terminate at Moreh. The total length of the highway in this region would be 762 km. (5) Kaladan Multi-Modal Transit Transport (KMTT) Project: This Rs. 545.83 crores project envisages connecting India's North-east with Sittwe port in Myanmar, thereby providing an alternate route for the transport of goods to the landlocked North-east. The project involves upgrading the Sittwe port and waterways in Kaladan river and the construction of 62 km of road from Kaletwa to the Indo-Myanmar border in Mizoram. The construction of this road is expected to be completed by 2011–2012. The project also involves the construction of a 117 km road on the Indian side from the Indo-Myanmar border up to NH-54. The handling points along the Indo-Myanmar border have been identified as Hmawngbu in Mizoram and Myeikwa in Myanmar. All these projects would no doubt improve both intra-region and cross-border connectivity of North-east. Improved connectivity in turn is expected to result in greater trade, thus bringing prosperity to the region. However, there are a few hurdles that impede the development of the road network in the region.

Challenges to the Development of the Road Network

Uneasy neighbourly relations

Strained relations with neighbours pose the greatest challenge in developing trans-border connectivity. This fact becomes amply clear when the trans-border road projects are examined. The fact remains that all the above-mentioned trans-border projects are focused on Myanmar, whereas the region also borders Bangladesh and China. Unfortunately, security implications *vis-à-vis* China and uncomfortable relations with Bangladesh have hampered the development of roads across the borders. It should be noted that the North-east region had traditional trading links with China, but following the 1962 border war, the borders were closed for trade. One of the various routes through which the region was connected with China was the 'Stilwell Road'. Following the opening of Nathu La in Sikkim for border trade, there has been a demand to open the Stilwell Road as well, for improving trade ties with China and Association of South East Asian Nations (ASEAN). China had also shown keen interest in its reopening, which would connect Yunnan province to Arunachal Pradesh *via* Myanmar. Although the Indian Government does not seem eager to reopen the road due to security considerations, it has sanctioned funds for upgrading the road from Ledo to Pangsau Pass (NH-153) to a two-lane road under SARDP-NE. At the same time, it has refused to upgrade this road in Myanmar (*The Assam Tribune*, 2008).

The development of trans-border connectivity between India and Bangladesh seems to be hostage to Dhaka's intransigence. Bangladesh has been the traditional market for the North-east and as mentioned earlier, before partition all transport routes from the region passed through it. Presently, trade with Bangladesh is carried through 12 Land Custom Stations (LCSs) located in the states of Meghalaya, Assam, and Tripura. A survey carried out to study the road conditions leading up to these LCSs on both sides revealed that the condition of most roads was poor.

Political decisions

Apart from the external factors, there are a few internal impediments to the development of the road network in the region.

The pattern of the road network in the region reflects, among other things, political decisions to construct roads in a particular area. These decisions, in turn, are influenced by various factors like faulty policies, vested interests, electoral compulsions etc. For instance, the decision for not building roads along the borders due to security considerations is a result of faulty policies. Thus today, the regions near the Sino-Indian and Indo-Myanmar borders are devoid of any roads. Similarly, the State Government's discrimination against building roads in the hills has led to the underdeveloped road networks in the North Cachar and Manipur hills in the states of Assam and Manipur. Another factor which influences decisions to develop roads is cost-benefit. The government seems reluctant to build roads in sparsely populated areas, arguing that the cost to build roads in these areas would far exceed the benefits it would accrue. This is one of the reasons for the lack of an extensive road network in Arunachal Pradesh.

Inefficient management

Inefficient management of resources, both at Central and State Government levels, has immensely contributed to the lopsided development of the road network in the North-east. Lack of funds, delay in acquisition of land, poor performance of contractors etc. are results of inefficient management, which acts as an impediment in building roads in the region. A paucity of funds as well as failure to disburse them in a timely manner has resulted not only in postponement of many projects, but also in their spilling over from one plan to the next. As a result, the monetary loss and resultant financial burden on the government is immense (*Annual Report 2006-07*).

Inordinate delays in the acquisition of land also contribute to the slow progress of the road construction projects. Elaborate procedural formalities and litigations lead to holding-up of the process of land acquisition. Further, since the entire region is ecologically fragile, the Environment and Forest Departments, both at Central and State levels, take a long time to clear road projects. Also, stringent state land Acts coupled with community ownership of land makes its acquisition extremely difficult. The non-cooperative attitude of various State departments further adds to the woes. In addition, shifting existing utilities like electric lines, sewer lines, water pipelines etc., and clearing the land for road construction takes a long time.

Road building in the region is also constrained by the unavailability of experienced contractors. Contractors from outside the region are reluctant to work in the North-east because of insurgency and low returns. And the local contractors lack the required expertise, resources, and experience to effectively manage these projects. Most of the construction agencies are owned by the local politicians who pressurise the government to grant projects to them. Thus, an unholy nexus between the local politicians, bureaucrats, and contractors is formed. They then indulge in corrupt practices such as using sub-standard materials, and not adhering to norms stipulated for building roads is quite common. In addition, non-availability of construction materials such as bitumen, stones, machinery etc. at the sites also contributes to slow progress in road building.

The BRO, which is the main agency for building roads in the North-east, is burdened with too many projects. Apart from building roads and bridges, it also builds other physical infrastructure in the border areas like schools, community centres, primary health centres etc. Also, in recent years the organization has been entrusted with building roads in the Naxal affected areas such as in Gadchirdi district. However, after the announcement of special packages for the North-east, additional units were sanctioned for BRO, but they are unable to shoulder the increased tasks.

Criminal-terrorism nexus

There is no doubt that insurgency in the North-east has degenerated into criminal syndicates. This growing criminal-terrorist nexus has emerged as a major hindrance not only to road construction but also to the overall development of the region. Potential investors shy away from this region because of insurgency. And the contractors and labourers who undertake these projects are deterred by extortion, kidnappings, and murder carried out by the rebels in the project sites. For example, on May 15, 2008, fearing attacks from rebels, more than 4,000 workers hired by 15 construction firms decided to withdraw from construction sites in Assam. Road building, thus, is adversely affected in insurgency affected states. Hence, the condition of roads in the insurgency-affected state of Manipur, where peace has reigned for some time, is pathetic.

Adverse terrain and climatic conditions

Building and maintenance of roads is extremely difficult because of the terrain and climatic conditions of the North-east. The region is mostly mountainous with moderate to steep gradient. These mountains are made of loose soil held tightly by the roots of trees and shrubs growing on it. Once the vegetation is removed during construction of roads, the mountains become destabilized. The situation is made worse by heavy rainfall, leading to landslides and floods, which not only makes building roads difficult, but also considerably damages the existing roads.

Recommendations

Several measures could be undertaken to overcome some of these impediments. For instance, the government's decision should be more influenced by the issue of social justice rather than cost-benefit factors. Its endeavour should be to provide adequate roads to the people residing in remote areas guided by the objective of balanced development of the region. Since the region is ecologically fragile, care should be taken for selecting the alignment of the proposed roads and proper scientific procedures should be applied while constructing roads. Proper surveys and detailed project reports should be meticulously prepared by involving all the concerned departments. To facilitate speedy approval of the projects, a coordinating body could be established to manage inter-departmental procedures. The establishment of an autonomous agency under the BRO to undertake all developmental road projects, other than strategic roads, could be envisaged. To accelerate road-building activities in the region, a public-private partnership model should be adopted by attracting the private sector along with foreign financial institutions like the World Bank and the Asian Development Bank for investing in the North-east must be made. Besides, there is a strong case for the restoration of the pre-partition road network with Bangladesh as it reduces trans-shipment cost and time between the North-east with the rest of the country. Therefore, India should endeavour to convince Bangladesh to provide a transit facility to India as well as jointly develop a road network along the borders. This would no doubt lead to mutual benefits for both the countries and the North-east in particular.

Conclusion

From the above discussion, it can be concluded that there has been no long-term comprehension that drove the development of roads in the North-east. Every one of the factors identified above has been a knee-jerk response to the internal and external developments. This is a consequence of a lack of understanding not only of the strategic importance of the North-east as a bridge that connects the rest of India with the East and South-east Asian neighbours. In addition, given several problems such as insurgency, underdevelopment, hilly terrain etc., what is needed is a long-term proactive approach to the defence and development of the region. And road communication is one of the means for securing defence and ensuring economic development.

References

Annual Report 1950–51, Ministry of Transport, Government of India, New Delhi, 1951, p. 9.

Report of Study Team on Administration of Union Territories and North-East Frontier Agency, Administrative Reform Commission, New Delhi, n.d., p. 42, as quoted in John W. Garver, "The Territorial Dispute", in *Protracted Contest, Sino-Indian Rivalry in the Twentieth Century*, University of Washington Press, Seattle, 2001, p. 99.

Chaube, Shibani Kinkar, Sunil Munsi, and Amalendu Guha, "Regional Development and the National Question in the North-east India", *Social Scientist*, 4(1), August 1975.

The Assam Accord of 1985, at http://www.satp.org/satporgtp/countries/india/states/assam/ documents/papers/assam_accord_1985.html (Accessed on April 22, 2017).

"Transforming the North-east", *High Level Commission Report to the Prime Minister*, Planning Commission, Government of India, March 7, 1997.

Pradhan Mantri Gram Sadak Yojana, at *http://pmgsy.nic.in/* (Accessed on May 29, 2017).

Raja Mohan, C., "Many Roads to Tibet Policy", Xinhua, August 3, 2006, at http://news. xinhuanet.com/english/2006-08/03/content_4915074. html (Accessed on April 25, 2017).

John W. Garver, "Development of China's Overland Transportation Links", *The China Quarterly*, 185, March 2006, pp. 11–12.

"Shared Integration", *Report of the Asian Dialogue Society's Greater Asia Initiative,* Information and Resource Centre, Singapore, 2006, p. 16.

Annual Report 2006–07, n. 15, p. 29.

Annual Report 2000–2001, Ministry of External Affairs, Government of India, 2001, p. 12.

"Construction of Trans Border Road to Begin This Month", Khonumthung News Group, March 4, 2006.

"India Gains Bigger Foothold in Myanmar", India Business News, New Delhi, October 27, 2007, at *http://www.iesingapore.gov.sg/wps/* (Accessed on April 15, 2017).

"Center Not to Upgrade Stilwell Road", The Assam Tribune, Guwahati, April 24, 2008.

19

Rural Road Connectivity and Poverty Alleviation in the State of Assam

Jitu Saikia and *Sumitra Bithi Kachari*

Introduction

Rural road plays a significant role in stimulating economic growth and social development. A well developed road connectivity ensures better agricultural income, facilitates the marketing of rural products, creates employment opportunities; and provides better access to healthcare, education etc., which enhance the efficiency and productivity of the rural people. Investments in infrastructure facilities have more lasting impact on reduction of poverty than unconditional subsidies and other financial assistance provided to the poor households. According to the Ministry of Rural Development (2015), Government of India (GoI), better roads provide improved market access, which results in favourable input and output prices leading to improvement in the economic condition of the rural poor. The significant justification of public investments on rural roads in India is to help the agrarian economy by a shift from traditional methods to modern cultivation practices. Apart from market access, improved information and extension services influence the cultivation choices and thereby, result in positive changes in cropping pattern.

Rural connectivity becomes a critical component in the socio-economic development of rural people. Two decades ago, there had been imbalanced development of the rural road network among the states in India. Some states provided sufficient road connectivity while some others did not have enough financial resources at their disposal and consequently, connectivity remained at low levels. Some of the major constraints faced by different states in providing road connectivity are insufficient funds for rural roads, inadequate maintenance, lack of proper specifications for quality upgradation etc. [Ministry of Rural Development (2011)]. With this backdrop, GoI had launched Pradhan Mantri Gram Sadak Yojana (PMGSY) in December 2000 to provide connectivity to unconnected eligible habitations and upgradation of existing roads to the standards. PMGSY has been implemented as a 100 per cent centrally-funded programme aimed at providing all-weather connectivity to all habitations of above 500 populations, and 250 in case of hills, desert and tribal areas of the country. The objective of PMGSY is to ensure sustainable poverty reduction. This chapter is an attempt to see the interlinking between road connectivity through PMGSY and rural poverty in the country with special reference to the State of Assam.

The objectives are to see how rural road connectivity affects poor households based on some existing literatures, and to investigate the interlinking between rural road and poverty by focusing on PMGSY.

Study Method and Data

The study is based on secondary information. Some existing literatures in the field of rural roads have been surveyed to find out the interlinking between rural road and poverty. Their findings have been considered in order to develop a link on how rural connectivity affects the poor households. Secondary data from sources, *viz.,* National Highway Authority of India, Ministry of Road Transport and Highways, Online Management, Monitoring and Accounting System (OMMAS) are collected in order to see the performance of PMGSY to the construction of rural roads in the country as well as in the State of Assam.

Rural Transport and Poverty

Historically, constructing roads has been associated with accessing resources and developing profitable markets with little attention to poverty issues. The poverty implications of transport investments were largely ignored as the sector was dominated by political, economic and engineering considerations (deGrassi, 2005). More recently, there have been several attempts from bilateral donors and multilateral development banks such as the World Bank, Asian Development Bank (ADB) etc., to review the evidence relating to rural transport and poverty (Starkey and Hine, 2014). The projects related to construction and development of roads, funded by ADB; reflect the pro-poor transport policies of these agencies.

Markets in developing economies are often characterized by spatial fragmentation due to poor transportation infrastructure. This inhibits households' and firms' ability to access goods and labour markets, technological innovations, and government services (World Bank, 2009). Policy-makers have increasingly attempted to address this problem by directing large sums of money towards the provision of roads and railroads. The primary channel through which road can affect rural households is *via* a reduction in transport cost. It facilitates trade in rural goods and services in particular, and increases the income of the poor households. A large number of literatures in international trade has established an inverse relationship between transport costs and trade flows (Baier and Bergstrand, 2001; Clark *et al.*, 2004; Feyrer, 2011; Storeygard, 2012). In a study Casaburi *et al.* (2013) find evidence that improved rural feeder roads facilitated easier market access for farmers, leading to a decrease in the observed market price of both rice and cassava. They have attributed this drop to a reduction in both transport costs and search costs.

Apart from the reduction in transport cost, rural road connectivity can potentially impact many other variables, such as access to credit, healthcare facilities, education, employment etc. After the construction of road, access to health facilities improves, and very substantial and critically important time can be saved. People are not only travelling outside to access health services, but also doctors and health workers are visiting these areas more frequently. Babinard and Roberts (2006) have reviewed several studies and found that poor road access was a major cause of pre-natal mortality. The study revealed that about 75 per cent mortality has resulted from inadequate transport to access basic health facilities and/or transport for referrals to other hospitals. Bell and van Dillen (2012) have examined the health benefits of villages connected by all-season roads through PMGSY scheme in Odisha. The data set and time period (short period) were not sufficient to identify statistically significant effects, but they clearly showed that sick people were increasingly taken to distant hospitals rather than being treated in village health centres. According to the Ministry of Rural Development Report (2015), better road connectivity is an important factor for improved coverage of child immunization during the last few years. It has been observed that there is an increase of 11 per cent children (five years or younger) in the habitations immunized as compared to the period before the road was constructed. Similarly, there has been an increase of 13 per cent in the percentage of children immunized during the same period in the control habitations also.

Education is another important element of development. Many important issues like, unemployment, health, sanitation, poverty etc., do get solved easily if people are educated. However, education in rural areas has its own challenges, which require different solutions. Among the different issues, poverty and poor connectivity are the main hindrances in achieving affordable quality education for all. Improved road connectivity has resulted in significant improvements in the education facilities in the villages and also access to higher education facilities available in the nearby urban cities in India (Ministry of Rural Development, 2015). The all-weather roads in the villages facilitate creation of infrastructure, institutions, increase in manpower and other improvements in the educational aspects. Different studies on correlation between road network and education also reveal the positive impact of road construction on educational services. McSweeney and Remy (2008) observed that maintaining rural pathways and feeder-roads in Peru, has increased primary school enrolment for girls by 7 per cent and secondary school attendance for boys by 10 per cent. In Nepal, the construction of a new rural road lead directly to the construction of a new secondary school funded by a private foundation (Starkey *et al.*, 2013). Mukherjee (2012) in an analysis of PMGSY road project, have found 22 per cent increase in school attendance as a result of the new village roads with enrolment from disadvantaged groups (STs and SCs) increased significantly. However, Aggarwal (2014) also identified a drop in enrolment for the age group of 14-20 years, which was thought to be related to greater employment opportunities for young people associated with improved road access.

India had started the era of planned development in 1951, and the government has been putting more emphasis on a framework for accelerated growth through investments in irrigation, power, heavy industry, transport etc. Besides, emphasis has been given on provision of social infrastructure and integrated rural development including agriculture. Rural roads act as a facilitator to the objectives of achieving sustain agricultural growth, improving basic health, accessing the schools etc., and thus hold the key to accelerated reduction of poverty. A study conducted by the Ministry of Rural Development, GoI in 2011 depicts the relationship between connectivity and poverty. According to the study report, states having low connectivity had higher poverty levels. Provision of good roads in rural areas also changes the characteristics of rural transport. With people tend to travel more, the ownership of vehicles increases. There is a shift from non-motorised vehicles to motorised ones and the cost and time of travel get reduced.

PMGSY and Rural Development in Assam

In India, a majority of the poor lives in rural areas. According to the report of the Planning Commission, GoI, about 25.70 per cent of the poor households lives in rural areas whereas the urban poor households accounts for only 13.70 per cent in 2011-12.[1] Lack of basic infrastructure facilities and transport make it difficult for them to access markets and other services. To address the issues of rural connectivity, the GoI has launched PMGSY in the year 2000. The fully centrally sponsored scheme covered a total of 1,78,184 habitations as per the criteria laid down. About 64 per cent of the eligible habitations (1,14,540) have roads today, which is a reasonable achievement for the programme. The scheme has been providing rural road connectivity of about 4,66,044 km, including upgradation of 1,67,977 km of existing roads since its inception.

---

1. The estimate is based on the poverty line given by Tendulkar Committee in 2009. The Committee defines rural poverty line as Rs. 27 per person per day and urban poverty as Rs. 33 per person per day. Recently, this has been revised to Rs. 32 per person per day in rural areas and Rs. 47 in urban areas by Rangarajan Committee.

The total expenditure incurred under PMGSY is Rs. 1,41,822 crore as on January, 2016 (Singh and Damodaran, 2016). The current sources of fund for PMGSY are cess on high speed diesel at a rate of Rs. 0.75 per litre, budgetary support, ADB funding, funding from World Bank and NABARD.

The rational of implementing PMGSY can be observed from the road network in India. The entire network can be categorized into four types such as National Highways and Expressways, State Highways, Major District Roads and Rural Roads.

Table 19.1: India's Road Network

Category of road	*Length (in km)*	*Percentage*
National Highways & Expressways	96460.72	2.88
State Highways	131899.00	3.94
Major District Roads	467763.00	13.98
Rural Roads	2650000.00	79.20
Total Length	**3346122.72**	**100**

Source: National Highway Authority of India.

Table 19.1 shows the percentage shares of these road categories to the total road network of the country. Among the four categories of road, about 79 per cent of the total road network in the country is occupied by Rural Roads. Therefore, better provision of these rural roads can provide a significant boost in economic activities of the agrarian economy.

In Assam, rural roads comprise over 82 per cent of the entire road network in the state. Table 19.2 shows the road connectivity in the state of Assam, showing a high concentration of rural roads in the state.

Table 19.2: Road Network in Assam in the Year 2013

Category of road	*Length (in km)*	*Percentage*
National Highways	2940	1.09
State Highways	3134	1.16
Other PWD Roads	42366	15.66
Rural Roads	222087	82.09
Total	**270527**	**100**

Source: Transport Research Wing, Ministry of Road Transport and Highways.

It is crucial to keep the roads serviceable throughout the year for agricultural growth, and access to other important services and facilities, *viz*. medical, education, market etc. The existence of poverty in rural areas of Assam is also high as compared with the national average as well as with the urban areas of the state. In the year 2011-12, the rural poverty in the state was about 34 per cent, whereas about 20 per cent of the population was below poverty line in urban areas. Lack of provision and maintenance of rural roads thus affects the poor people severely.

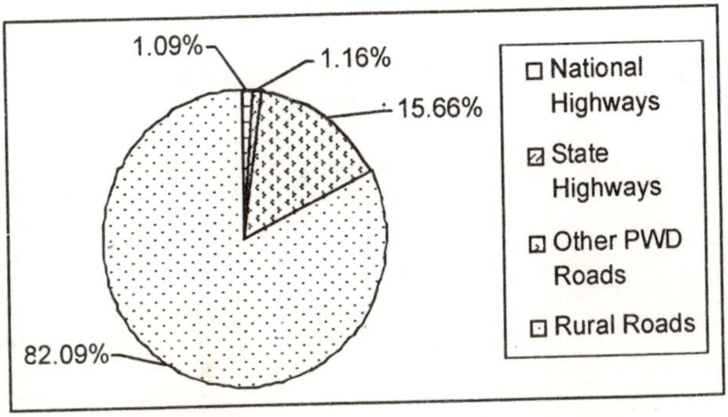

Fig. 19.1: Road Network in Assam.
Source: Ministry of Road Transport and Highways.

Lack of all weather roads in rural areas increases the time requirement to access other social infrastructure and market. The time factor is most crucial in case of accessing rural health services in general and, maternal and child health services in particular. A significant number of deaths of women and children in rural areas can be attributed to poor quality of road in rural areas of the state. Thus, rural development has become an important matter for economic upliftment, social justice and inclusive growth in the state. The provision of rural road is a key component to rural development, which enable the rural people to have access to schools, health centres and markets. Rural roads serve as an entry point for poverty alleviation because lack of access is accepted as a primary cause of poverty (Ministry of Rural Development, 2011).

Table 19.3 shows the achievements of PMGSY to provide rural connectivity in the state of Assam.

The total length of rural road targeted under PMGSY in Assam was 17,098 km up to the year 2015-16. The success of the programme in the state is significant as about 93 per cent (15,981.96 km) of the total targeted length has been completed in the period. Apart from that 8,163 villages of the state have connected through the scheme up to the year 2015-16. This is about 32 per cent of the total rural habitations in the state. PMGSY has made it possible for producers of perishable commodities such as milk, fish and vegetables to sell their products to a wider base of consumers. Equally, it has enabled companies to distribute their products through rural retail stores (Singh and Damodaran, 2016).

Table 19.3: Rural Road and Habitation Connectivity under PMGSY in Assam up to the Year 2015-16

Year	Target length	Completed length	Target Habitation	Completed Habitation
2005-06	1037	459.37	720	250
2006-07	2474	884.24	1300	501
2007-08	1685	772.74	995	410
2008-09	2730	1328.73	1860	699
2009-10	2585	1555.38	1350	912
2010-11	2008	1628.64	400	817
2011-12	1224	2564.98	500	1360
2012-13	1175	2171.06	237	1014
2013-14	650	1978.62	160	994
2014-15	720	1447.09	228	652
2015-16	810	1191.11	338	554
Total	**17098**	**15981.96**	**8088**	**8163**

Source: PMGSY, Online Management, Monitoring and Accounting System (OMMAS).

However, the implementation and coverage of PMGSY in the state is low as compared with other states of the country. The states that have recorded the highest road construction under PMGSY are Madhya Pradesh (63,548 km), Rajasthan (58,462 km), Uttar Pradesh (45,905 km), Bihar (35,510 km), and Odisha (35,019 km). The higher allocations of fund to these states may be due to the selection criterion of the scheme, which basically puts more weightage on population rather than remoteness of a particular area. Under PMGSY, state-wise allocations are fixed based on population criteria, which automatically ensures more money being released and road getting built in Bihar and those other states having more population density rather than the North-Eastern States having lower population density.

Conclusion

PMGSY can provide solutions for more effective long term poverty alleviation in the State of Assam. It helps to raise overall productivity and expansion of non-farm employment opportunities in the state. As the incidence of poverty is more in rural areas of the state as compared with the urban areas, better road connectivity of rural areas ensures more inclusive development of the state. A higher rate of completion of work under PMGSY indicates that the programme is less prone to leakage, which is, otherwise, a common feature of most of the poverty alleviation programmes implemented in the state. As the state economy is dominated by rural sectors, it is important to develop the rural areas by providing basic infrastructural facilities. As an infrastructural development programme, PMGSY will have long term impact on the development of the state.

References

Aggarwal, S. (2014): "Do Rural Roads Create Pathways Out of Poverty? Evidence from India", *Job Market Paper,* Santa Cruz University of California.

Babinard, J. and P. Roberts (2006): "Maternal and Child Mortality Development Goals: What can the Transport Sector Do?", *Transport Paper -12,* Washington DC, USA World Bank.

Baier, S.L. and J.H. Bergstrand (2001): "The Growth of World Trade: Tariffs, Transport Costs, and Income Similarity", *Journal of International Economics,* 53(1): 1- 27.

Bell, C. and S. van Dillen (2012): "How does India's Rural Roads Programme Affect the Grassroots? Findings from a Survey in Orissa", *Policy Research Working Paper 6167,* Washington DC: USA, World Bank.

Casaburi, L.R. Glennerster, and T. Suri (2013): "Rural Roads and Intermediated Trade: Regression Discontinuity Evidence from Sierra Leone", *Working Paper,* MIT.

Clark, X.D. Dollar, and A. Micco (2004): "Port Efficiency, Maritime Transport Costs, and Bilateral Trade", *Journal of Development Economics,* 75(2): 417-450.

deGrassi, A. (2005): "Transport, poverty and agrarian change in Africa: models, mechanisms and new ways forward", *IDS Bulletin,* 36(2): 52-57.

Feyrer, J. (2011): "Distance, Trade, and Income—The 1967 to 1975 Closing of the Suez Canal as a Natural Experiment", *Meeting Papers No. 1438,* Society for Economic Dynamics.

McSweeney, C. and M. Remy (2008): "Building Roads to Democracy? The Contribution of the Peru Rural Roads Programme to Participation and Civic Engagement in Rural Peru", *Social Development Notes 111,* Washington DC, USA: World Bank.

Ministry of Rural Development (2015): "Impact Assessment Study of Improved Rural Road Maintenance System under PMGSY", Government of India, October.

Ministry of Rural Development (2011): "Final Report of the Working Group on Rural Roads in the Twelfth Five Year Plan", Planning Commission, Government of India, October.

Mukherjee, M. (2012): "Do Better Roads Increase School Enrolment? Evidence from a Unique Road Policy in India", Seminar paper posted on Social Science Research Network.

Singh, S. and H. Damodaran (2016): "Pradhan Mantri Gram Sadak Yojana: How the Programme Impacted Indian Hinterland", *The Indian Express,* updated in February, 25.

Starkey, P.A. Tumbahangfe and S. Sharma (2013): "External Review of the District Roads Support Programme (DRSP) Final Report", Swiss Agency for Development and Cooperation, Kathmandu, Nepal.

Starkey, P. and J. Hine (2014): "Poverty and Sustainable Transport: How Transport Affects Poor People with Policy Implications for Poverty Reduction", UN-Habitat and the Overseas Development Institute (ODI), October.

Storeygard, A. (2012): "Farther on down the Road: Transport Costs, Trade and Urban Growth in Sub-Saharan Africa", *Working Paper,* Tufts University.

20

Significance of Communication in Cultural Interaction and Traditional Economic Organization

A Study on the Rural *Hats* of Karbi Anglong, Assam

Jushna Beypi and *Rengmini Tokbipi*

Introduction

The Anthropological study of economic organization is a focal point for discovering the relationship between material and non-material aspects of the environment and human behaviour as well as the social system within which it is exchanged and redistributed as cultural goods and services. The economic organization is treated as a link between the material base to culture and social structure. Economic order in one form or another is a functional prerequisite to societal survival and continuity (Hoebel-Frost,1976:247). From the anthropological point of view economy is believed as a system of managing the production, distribution and consumption of goods and services in a society. There is also stress on the fact to the motivations for producing, exchanging and consuming goods and services which are shaped by cultural traditions (Dash, 2004:123). Piddington (1952) says "economic system is designated to satisfy material wants of the people to organise production, to control distribution and to determine the rights and claims of ownership within the community" (*see* Dash, 2004:123). Every society has access to natural resources—land, water, plants, animals, minerals and every society has cultural rules for determining who has access to particular resources and what can be done with them (Ember, 2002:286). The natural environment is the origin of all the resources people use to secure the material needs and desires. However, societies have only limited access of resources offered by nature. These limitations are imposed by where the people live, who they trade with and the technology at their disposal. Moreover, no society uses all available elements in the environment. All groups are selective choosing to utilize certain resources and to ignore others (Hunter-Whitten, 1976:237). All societies have customs specifying access to natural resources; customary ways of transforming or converting those resources through labour into necessities and other desired goods and services; and customs for distributing and perhaps exchanging goods and services (Ember, 2002:305). In this regard every society makes use of a technology, which includes tools, constructions and required skills. The technology and material culture of a people are so intertwined with their economic life that they can be treated as separate only conceptually (Clifton, 1968:187).

Each local economic system represents a more or less unique aggregation of experience built-up partly through progressive innovation by the people themselves as they have adjusted to their local setting. Actually, the economic system of any people quickly reveals that behaviours relating to production, exchange, property and consumption all tie in functionally with social organization and with the political, legal, religious and aesthetic behaviours (Kessing, 1958:223). The economic behaviour of the people lies in their own cultural terms, focusing on the substantive social and technological processes through which human interact with one another and with the environment to meet their subsistence needs (Hunter-Whitten, 1976: 251). The study of the relations between cultural patterns and physical conditions is of the greatest importance for an understanding of human society and it is intimately co-related with the economic organization of the concerned society.

Present Study and Studied People

The present study has been done in the Karbi Anglong district of Assam. Karbi Anglong is one of the two hill districts of Assam, has two parts that is East and West Karbi Anglong.

It is situated in the central part of Assam. It is bounded by Golaghat district in the east, Meghalaya and Morigaon district in the west, Nagaon and Golaghat district in the north and N.C. Hills district and Nagaland in the south. The district with dense tropical forest covered hills and flat plains is situated between 25°33′N to 26°35′N Latitude and 92°10' E to 93°50' E Longitude. It is the Largest district of Assam with a total geographical area of 10, 434 sq. kilometer (KAAC, 2016). The Karbis were the earliest settlers and numerically dominant Tribal group of the said region. Different other Tribal groups like Dimasa, Bodo, Naga, Hmar, Khasi, Kuki, Hajong, Man (Tai) etc. and different other communities were also dwelling in the studied region. The present study has been conducted in different traditional tribal markets or *Hats* organized in different places of Karbi Anglong district (Census, 2011).

Theoretical background of the study: The present study is based on the theory of functionalism and the concept of need as proposed by Malinowski, (1922). It has been tried to analyze from the perspective of Anthropology of communication.

Aims and Objectives

The present study is primarily focussed on three important aspects—

1. To study about the traditional *Hats* of the Karbi Anglong district.
2. To study about the multi-ethnicity interaction of different cultures.
3. To study about the role of roads in terms of communication and economy of the Karbi people.

Methodology Applied

The study was primarily based on qualitative data to collect information from the multi-ethnicity community regarding their significance of communication in cultural interaction and traditional economic organization. In view of the present study observation technique was applied for an in-depth knowledge about the traditional weekly markets and their economic perspectives. Data were collected on several belief systems and practices associated with their traditional economic organization. The study was conducted during the period of January 2016 to December 2016. There were four divisions of the total field-work.

Division-1: Foremostly, the traditional markets or *hats* in different sub-divisions and villages were visited. Exchange of different products were observed and interaction was done with the people of different ethnic groups who assimilated there.

Division-2: Case studies were taken on the concerned people associated with the selling of different products in the traditional markets or *hats.* It was focused to know about their traditional number systems as practiced in their traditional markets.

Division-3: Detailed open structured interviews were taken from the key informants and from the people of different ethnic groups who were the regular dependants on their traditional markets, for the sustenance of their livelihood. It was focused to find out the socio-economic and socio-cultural relevance of traditional markets or *hats.*

Division-4: Data were collected about the barter system of the studied people.

Result and Discussion

A. Rural Hats of East Karbi Anglong

East Karbi Anglong is the major district of Assam and Diphu is the Headquarter of it. Through the National Highways 36 NH and 39 NH; railway station Diphu is well connected with several other districts of Assam like Silchar, Guwahati, Nagaon, Tezpur, Golaghat, Jorhat etc. and other states like Nagaland and Meghalaya. A number of traditional weekly markets or *hats* were found within Diphu town at *Larulangso* and in several other places like at *Manja, Delai, Desobai, Bakulia, Phuloni, Langhin, Dokmuka, Dangoan, Parokhua* market. These weekly traditional markets were communicated through NH-36. On the other hand, NH-39 was connecting the traditional weekly markets or *hats* conducted at *Upper-Deopani, Deopani, Khatkoti, Sarihojan, Silonijan, Japrajan, Balipathar, Daithor*. People of different tribal groups like Karbi, Dimasa, Boro, Rengma Naga and other communities like caste Hindu groups, Muslims, Nepalese were regularly coming to those traditional weekly markets from different places and purchasing different organic resources for daily livelihood in cheaper prices from local people. It was observed that in between the Karbi people even barter exchange method was being practiced. In certain cases barter exchange has been noticed in between the Karbi people with the Boro and Dimasa people as well. However, with the caste Hindu people and Muslims always the monetary exchange has been practiced. It was observed that, particularly the caste Hindu and Muslim people used to come to such traditional markets from outside areas with different products like second hand clothes, spices, some attractive items etc. The local people had a scope to purchase them within their close vicinity, which used to save their cost of transportation instead of going to the markets of town as well as the prices of such objects were much lesser in the *Hats,* if compared with the regular markets of the town. Here the communication and transportation used to play a vital role in cultural interaction.

Different Economic Resources: It can be noted that the traditional occupation of the Karbi people was shifting hill or *Jhum* cultivation and on the other hand they used to collect forest resources for consumption and selling. Along with that they were associated with traditional weaving, handicrafts, kitchen gardening and animal husbandry. The traditional occupation of the Dimasa people was also the shifting hill or *Jhum* cultivation. Along with that they practiced traditional weaving, kitchen gardening etc. The Bodos were the settled agriculturist mainly. The Nagas were traditionally associated with Terrace and *Jhum* cultivation. It can be noted that, geographically Karbi Anglong is recognized as a hill district and fertile soil is less prevalent in the studied area. On the other hand, it comes under rain shadow region. Thus, the level or quantity of agricultural production was quite lesser. In this regard, the Karbi people mostly produced paddy, potato, maize through *jhum* cultivation; the Nagas were the prime producers of paddy, pineapple, potato through *jhum* and terrace cultivation; the Bodos were the producers of paddy through settled cultivation and the Dimasas were primarily engaged with *jhum* cultivation to produce ginger, paddy. Different vegetables, leafy vegetables and fruits were primarily produced through kitchen garden. Fishes and other water bodies were collected by them through local streams only. However being the rain shadow region, particularly during summer and winter, the level of water used to get much decreased.

In the studied region although different types of productions were available, but the lesser quantity of such produces, could hardly satisfy the different requirements for the individual families, related to their food and other essential livelihood needs. In this regard, ultimately they had to depend and share the productions of each other for their sustenance; in such cases, the traditional weekly markets are the prime platform.

The local Karbi people came down from the hills to sell their organic resources in different traditional weekly markets. They used to carry various local produces, vegetables and natural forest resources. Those were like pumpkin, yam leaf, yam, banana, sugarcane, papaya, chilli, betel nut, betel leaf, banana flower, brinjal, bamboo shoot, mashroom, pumpkin leaf, different local leafy vegetables like *Hanrisang, Hansangbi, Hanbipo, Torbob, Dumkek, Tara, Hanjang, Hanthu, Theso-kumbong, Theso-keho, Aantimi, Jokaan, Hendonngdar, Hanserong, Pherklung, Chushot, Hantharvo, Hanche* etc.

Table 20.1: Traditional weekly markets on or communicated through NH 39

Name of the traditional markets	*People coming from*	*Distance*	*Transportation*	*Conducted on*	*Language or dialect communicated*
Balipathar market	Dimapur, Bokajan, Borpathar, Sorupathar	4 km	Tempo, Carrier van, Auto Truck	Every Monday	Karbi, Assamese, Nagamese, Nepali
Upper Deopani market	Bokajan, Borpathar, Silonijan, Golaghat district, Jorhat	5 km	Auto, Tempo, Carrier van, Auto truck	Every Wednesday	Karbi, Assamese
Deopani, Khotkhoti market	Dimapur, Bakajan, Golaghat district, Jorhat, Tinsokia, Borpathar	10 km 7 km	Truck, Tempo, Carrier van	Every Thursday	Karbi, Assamese, Nagamese, Nepali
Sarihojan, Silonijan market	Bokajan, Dimapur, Manipur, Golaghat district, Jorhat, Tinsokia	6 km 11 km	Bus, Carrier van, Auto Truck, Truck	Every Friday	Karbi, Assamese, Nagamese,
Japrajan market	Bokajan, Deithor, Borpathar, Sorupathar, Golaghat district	6 km	Truck, Bus, Carrier van, Tempo	Every Saturday	Karbi, Assamese

Source: Authors own data collected from field.

Further, it was noticed that in the traditional *Hats*, different vegetables, leaves and fruits were sold by the females whereas, the males were dealing with paddy, fire wood, fishes.

Table 20.2: Traditional weekly markets on or communicated through NH 36

Name of the traditional markets	*People coming from*	*Distance*	*Transportation*	*Conducted on*	*Language or dialect communicated*
Langhin market	Nogaon, Doboka	13 km	Auto, Tempo, Auto Truck, Truck, Carrier van, Bus	Every Sunday	Karbi, Bodo, Assamese, Hindi
Centre market	Hawraghat, Bakalia	9 km	Bus, Tempo, Truck, Carrier van	Every Monday	Karbi, Assamese, Bodó
Bakulia, Dangaon market	Nagaon, Doboka, Dokmuka, Kothiatuli,	37 km 10 km	Auto, Tempo, Auto Truck, Truck, Carrier van, bus	Every Tuesday	Karbi, Assamese
Dokmoka market	Doboka	12 km	Truck, Bus, Carrier van, Tempo	Every Wednesday	Karbi, Assamese, Bodo
Phuloni market	Dokmuka, Bakulia	8 km	Bus, Truck, Tempo	Every Thursday	Karbi, Assamese, Bodo
Manja market	Dimapur, Diphu	16 km	Carrier van, Truck, Bus, Tempo, Auto truck	Every Friday	Karbi, Kuki, Assamese, Nagamese, Hindi
Dilai, Parokhua market	Dimapur. Kothiatuli, Kampur, Nogaon, Doboka	33 km 93 km.	Bus, Truck, Auto truck, Tempo	Every Saturday	Karbi, Assamese, Nagamese

Source: Authors own data collected from field.

B. Traditional Weekly Markets or Rural *Hats* of West Karbi Anglong

The West Karbi Anglong district is a new proposed district formed out of the existing Karbi Anglong District of Assam in 2016. Hamren is the headquarter of the newly formed district. The district is a part of Karbi Anglong Autonomous Council and it is administered according to the Sixth Schedule of the Indian Constitution.

Karbi language is widely spoken in the district. English, Assamese and Hindi are the other languages that are spoken in the district. The district is inhabited by many tribes, Karbi being the majority, along with Tiwa, Garo, Bodo, Dimasa tribal groups and caste Hindu, Muslim communities.

The district headquarter is connected by road. Karbi Anglong Autonomous Council Transport buses ply at regular interval from the district headquarters to important places like Guwahati, Diphu, Lanka and Hojai. Also one Meghalaya Transport runs regularly to and from Hamren to Jowai. The nearest railway station is at Hojai, which is around 47 km from Hamren. The road that was connecting the studied regions became beneficiary for the local people in terms of transport, communication and for their economic upliftment.

In this west part of Karbi Anglong National Highway 27, connected Neilie to Baithalangso and was connected to Kheroni crossing. Another local road was connecting Baithalangso to Donkamokam to Kolonga to Jengkha to Kheroni crossing and this was ultimately connected to Lanka and then to N.H. 27. Again, from N.H. 27, West Karbi Anglong was connected through PWD road to Tumpreng to Hojai to Nilbagan and to N.H. 27. Tumpreng was at the border of Karbi Anglong separated through Kopili river from Hojai. That road ultimately connected Hojai to Karbi Anglong passing through the important areas of West Karbi Anglong. It lead to headquarter Hamren and further got connected to Meghalaya.

Economic Resources

The tribal communities of the studied region were interacting frequently in terms of economy, various goods and services, in their traditional weekly markets. Many of the tribal and non-tribal communities could be seen to interact in their traditional markets during the marketization of different productive goods. West Karbi Anglong was the rich zone for the cultivation of various organic resources. The Karbis of those region cultivated organic ginger; broom grass, pineapple, pumpkin, sugarcane to produce jaggery, pear and organic chillies of different kinds. Apart from this, the Karbi women collected organic herbs and ferns to sell in the traditional weekly markets. Along with that, they used to sell ferns and other natural resources such as bamboo shoot, and different local vegetables like *Inglong-a-bobok, inglong-a-behendi, dungkek, bonghom arvo, langdung, tara, henru, hanserong, hippi, hen-up, rui pharkong, plimplam, thichi, ginger, ahan-a-lothe, kolmuk, lang-a-churu, ingki ok, pijo alang, hen dongdar, chosot, hanthu, papaya, birik man, birik losa, birik jangkek, dido, mula arvo, musi, jok-an, pharklung, hantharvo, hen soksu, henjor, hen kove, inglong kenong ahen, mu-plakplak, mu-arnan, mu-rik, mu-plong, henru, nempo arvo, mora arvo, phi-kangnek, bonghom arvo, mehek, churu, langdun, kove* (betel nut), bikron (betel leave), *phinu* (banana), *ahan a lothe* (banana used for cooking) etc. Moreover, different handicrafts like *tar* (mat made-up of bamboo), *beleng, hoton* (a kind of basket made-up of bamboo and cane) were also sold. *Theng* (firewood) was usually sold by male.

Tumpreng, Donkamokam, Hamren, Kolonga, Jengkha, Kheroni and Baithalangso were the prime markets where most cultural interaction used to take place. Yeats were mainly sold in Donkamokam weekly market; broom grass and organic ginger were the main source of market in Hamren; in Tumpreng market other than selling of various vegetables and market products, animals like cow, piglet, goats, duck and hen and cocks were the main sources of marketization. While in the weekly market of Kolonga, Jengkha and Kheroni, organic ginger, organic chilli, broom grass, pineapple, pear and jaggery are the main sources. However such sources became variable according to seasonal production. Mostly different organic products are being supplied from the nearby hilly areas of Jirikindeng.

In those weekly markets traditional ethnic food stalls could also be noticed, mostly ran by the Karbi women, in those food stalls traditional country liquor like Hor-arak, Hor-lang and Hor-po were also available. In some of the places like Kheroni and Tumpreng, a few food stalls were run by the non-Karbi women with other food items. The people of other ethnic groups coming from the neighbouring districts and state, used to bring different objects like utensils, dry fish, different vegetables like potato, tomato, onions, garlic etc., cloths, different kinds of iron implements like knives, variety of spices, variety of colourful yarns etc.

The traditional weekly markets of the studied areas were the prime place of interaction between different ethnic groups like Bodo, Garo, Karbi, Mizo, Khasi, Hmar, Panar, Kuki, Manipuri, Dimasa and other ethnic groups like Bengali, Assamese, Bihari and Muslims. Such interaction played a vital role in their economic life as well as social life.

Table 20.3: Traditional weekly markets to be communicated through NH-27

Name of the traditional markets	*People coming from*	*Conducted on*	*Language or dialect communicated*	*Distance*	*Transport*
Tumpreng	Hojai	Every Monday	Karbi, Garo, Dimasa, Assamese, Bengali	14 kms. (Hojai)	Bus, Auto, Truck, Tempo etc.
Baithalangso	Hojai, Donkamokam, Hamren.	Every Tuesday	Karbi, Garo, Dimasa, Assamese, Bengali	34 kms., 26.8 kms., 26.7 kms., 28.3 kms (SH 18) respectively	Bus, Auto, Truck, Cycle, Tempo etc.
Jengkha	Lanka, Hojai, Jirikindeng	Every Wednesday	Karbi, Garo, Dimasa, Bodo, Assamese, Bengali	18 km. (Lanka),	Bus, Auto, Truck, Tempo etc.
Dongkamokam	Hojai, Tumpreng, Baithalangso, Hamren, Kolonga	Every Thursday	Karbi, Garo, Dimasa, Bodo, Assamese, Bengali, Garo	22 km. (Hojai)	Bus, Auto, Truck, Tempo etc.
Kheroni	Hojai, Lanka	Every Friday	Karbi, Assamese, Bengali	28 km. (Hojai), 14 km. (Lanka)	Bus, Auto, Truck, Cycle, etc.
Hamren	Liumphung, Linchika, Rumphung, Rongmandu, Socheng, Taradubi, Umcherra, Hojai	Every Saturday	Karbi, Khasi, Mizo, Bodo, Hmar, Panar, Kuki, Manipuri, Assamese, Bengali	Rumphung 38 km, Langsomepi 8 km, Longchek 24 km, Birmaine 30 km, Rongmandu 10 km, Umcherra 16 km, Borpu 9 km, Somupong 4 km, Langmet 6 km, Harlonglangdo 12 km, Tirkim 7 km, Taradubi, Khanduli 35 km, Am-e 32 km, Langhemphi 30 km,	Bus, Auto, Truck, etc.

Table 20.3—(*Contd.*)

Name of the traditional markets	*People coming from*	*Conducted on*	*Language or dialect communicated*	*Distance*	*Transport*
				Tapat 22 km, Rongpangbong 27 km, Mukoiram 34 km, Khilimlong 31 km, Socheng 28 km, Langerdang 43 km, Tadolangso 30 km, Nartyang 65 km, Kaisama 60 km. Hojai, Baithalangso. To Hamren	
Borthol/Kolonga	Hamren, Donkamokam, Tumpreng, Baithalangso, Hojai	Every Sunday	Karbi, Dimasa, Assamese, Bengali.	Lanka *via* Kheroni-Jengkha-Dongkamukam with distance around 35 km.	Bus, Auto, Truck, Cycle, Tempo etc.

Source: Field Survey.

Therefore, these roads are the vital part of Communication, Culture, Economy, Multi-ethnicity interaction between the neighbouring districts for local tribal people residing in the West Karbi Anglong District of Assam. In this regard, the people from the neighbouring districts and states buy the local organic resources in cheap prices and supply in the other nearby town areas selling them in to higher prices in different towns.

Concluding Remarks

Geographically, Karbi Anglong comes under the rain shadow region, where the annual rainfall is lesser than the other regions of Assam. On the other hand, the traditional occupational pursuit like *jhum* cultivation is based on simple technology and indigenous method. On the other hand even the settled cultivation is not supported by irrigation anywhere in the studied areas. Thus, the level of production is not very high, which can satisfy the adequate requirement of food as well as other common minimum livelihood requirement of the studied families. In this regard, the sharing of indigenous productions in between the Karbi people as well as with the other community members is the only possible way to sustain their livelihood. In this regard, the traditional weekly markets or the tribal *hats* are the prime platform of economic transaction. Economic requirements as well as several other livelihood requirements have played a vital role to integrate different ethnic groups in spite of their cultural differences. In this regard, the surface communication and transportation has played the prime essential role to adjoin the remote corners of the district as well as the interior regions of their settlement. Even it has brought the indigenous people of neighbourhood districts and states in the common economic sharing. Thus, it can be summarized that in the studied region surface communication has played a prime role in the sharing of economic resources, for livelihood sustenance as well as cultural sustainability in the studied region.

Acknowledgement: The authors sincerely acknowledges to the Honourable Pro-Vice Chancellor Sir of Assam University, Diphu Campus; Head, Department of Anthropology, AUDC; Research Supervsior and all of the studied people for their kind academic support always. The authors are extremely grateful to Dr. Tarun

Dutta Sir of the Department of History, AUDC for his kind consideration of this paper for presentation in the seminar and its further proceedings.

References

Census of India, 2011. New Delhi: Government of India.

Dash, K.N., 2004. *Invitation to Social and Cultural Anthropology*. New Delhi: Atlantic Publishers.

Ember, C., with M. Ember and P.N. Peregrine, 2002, *Anthropology*. New Delhi: Pearson Education Asia.

Hoebel, E.A., and E.L. Frost, 1976, *Cultural and Social Anthropology*. New York: MacGraw Hill Book Company.

Hunter, D. E., and P. Whitten, 1976, *The Study of Anthropology*. New York: Harper Row Publishers.

Kessing, F.M., 1958, *Cultural Anthropology: The Science of Custom*. New York: Holt Rinehart Winston.

Official Records of Karbi Anglong Autonomous Council, 2016. Diphu: Karbi Anglong.

21

Impact of Road Communication on Agricultural Marketing with Special Reference to Karbi Anglong District of Assam

Mintu Tokbipi and *Juli Tokbipi*

Introduction

Transport and Communication is the vital infrastructure of a modern economy. An efficient transport system integrates remote, backward and urban area and plays a vital role in increasing productivity and improving quality of life. Moreover, development of this sector generated large employment opportunities. Karbi anglong district is well-connected with other districts through various routes. On the North, it is covered by NH-37 and on the East by NH-39. The internal route of the district is under Public Works Department (PWD).

In the Karbi Anglong District about 65 per cent of the total area is hilly and more than 85 per cent of the rural population directly depends on agriculture for their livelihood. In the Karbi Anglong district main agricultural practice is the *Jhum* Cultivation. But due to the absence of proper road transportation facilities in the rural areas, village farmers cannot reach nearby market (*mandis*) to sell their produce at a fair price. Thus, they prefer to sell their produce at the village markets itself. Agricultural marketing system is an efficient way by which the farmers can dispose their surplus produce at a fair and reasonable price. Improvement in the condition of farmers and their agriculture depends to a large extent on the elaborate arrangement of agricultural marketing. The term agricultural marketing include all those activities which are mostly related to the procurement, grading, storing, transporting and selling of the agricultural produce. Thus, agricultural marketing includes the operations like collecting, grading, processing, preserving, transportation and financing.

Area of the Study

Karbi Anglong District is a Six Schedule Constitutional District of Government of India. Karbi Anglong district aspires the status of an autonomous hills district under para-1 having independent district council under para-2 with the headquarters at Diphu. And under para-20 of the same schedule, Karbi Anglong is declared as a tribal area which came into being on 23rd June, 1952 while Late Bishnu Ram Medhi was the Chief Minister of Assam at that time. The district covers an area of 10,434 sq.km. (Rural: 10,396.55 sq.km. and Urban: 37.45 sq.km.) out of the Assam total area of 78,438 sq.km. The District is divided into 4 Revenue Circles which cover 2,921 villages. In the district total population is 9,56,313 out of which 8,43,347 rural population and 112966 urban population. The district has a tropical climate in general and specifically, the entire district experiences different climate in different part due to wide variation in the topography. The monsoon starts generally from mid-May and continues till September. Rainfall is not uniform throughout the district. Though the average rainfall in the district is 1147 mm, but it is not uniform throughout the district, therefore, there is a wide range of disparity in distribution of rainfall from place to place. Kheroni and Amreng areas of Hamren sub-division and Dhansiri area of Diphu sub-division are known as rain shadow area. The area between Kheroni to Dhansiri receives less than 100 mm. rainfall. During summer, the atmosphere becomes sultry.

Objective

1. Agricultural marketing system in the District.
2. Impact of Road transportation on Karbi Anglong Agriculture.
3. Solution for the development of Agriculture sector.

Methodology

The present study is based on both primary and secondary data. Secondary data collected from Assam Economic Survey, article published in various journals and Magazines and newspaper, website. The primary data were collected from the village producers and sellers through direct personal interview method.

Agricultural marketing system in the District

Under the existing system of agricultural marketing in Karbi Anglong District, farm produces change many hands before reaching the consumer. The middleman, in the process extracts a large share of consumer's rupee while the farmer gets only a small share. Taking advantage of the farmer's poor condition and weak bargaining power in the transaction, a set of middleman at different stages have emerged as a powerful channel for the procurement of food grains. They by and large, dictate the price. Thus, the farmer needs protection from such exploitation and also assistance in many respects, more so in a State like Assam where the agriculturists are subject to natural calamities and various other constraints.

The existence of an efficient marketing system is an important prerequisite for the better performance of agriculture. Because it is through an efficient marketing system the farmers get remunerative prices for their products and they are encouraged to produce surplus for sale. But unfortunately the marketing infrastructure is not properly developed in the district. Still a major part of the marketed surplus of the farmer in usually sold in the village market called "Hats", which are generally running either on congested space without having infrastructural facilities or on National Highways or State road. Under this type of free marketing system, the small farmers do not get reasonable prices for their crops because the marketed produce of the farmers move from rural markets to the large secondary markets in the urban areas through a number of middlemen, who generally indulge in malpractices. Higher price paid by the consumers to the middlemen do not reach the actual produces.

In Karbi Anglong District four different systems of agricultural marketing are prevalent:

- **Sale in Village:** The first method open to the farmers in

Karbi Anglong District is to sell their surplus produce to the village money-lenders and traders at a very low price. The money-lender and traders may buy independently or work as an agent of a bigger merchant of the nearby *mandis*.

- **Sale in Markets:** The second method of disposing surplus of the Karbi Anglong farmers is to sell their produce in the weekly village markets popularly known as *'hat'* or in annual fairs.
- **Sale in *Mandis*:** The third form of agricultural marketing in Karbi Anglong farmers is to sell the surplus produce through *mandis* located in various small and large towns.
- **Co-operative Marketing:** The fourth form of marketing is the co-operative marketing where marketing societies are formed by farmers to sell the output collectively to take the advantage of collective bargaining for obtaining a better price.

Impact of Road Transportation on Karbi Anglong Agriculture

Road Communication plays an important role in the district for agricultural marketing of goods and services. With the help of a better road communication facilities village farmers can easily sell their surplus produce in the market without the help of the middleman. For that they are getting actual price of their produce. But, still there are some villages in the district which are not well-connected with the main roads or towns for proper functioning of the whole year. For that farmers have to sell their surplus produce at the village level to middleman at very low rate. At this rate their cost of production will be more than their selling price. For that buyers have to pay a higher price of their consumption goods. For example when the farmers sell Ginger in the village-level, they can get only 30 to 40 Rupees per kg., but if they can sell in the proper market, they can easily get 100 to 140 Rupees per kg. All these because of not proper communication facilities in the village-level.

Problems of Road Transportation on Agricultural Marketing in Karbi Anglong District

Road transportation system in Karbi Anglong District is suffering from numerous problems—

1. The road transport system in the district is suffered from unfavourable road conditions. Existing roads in the district are not at all conducive to operate a smooth and efficient road transport system. Moreover, a large number of rural areas do not have any road linkage. In the absence of all weather roads, road transport system is facing a setback condition particularly during monsoons.
2. In the absence of proper road transportation facilities in the rural areas, Karbi Anglong district farmers cannot reach nearby *mandis* (markets) to sell their produce at a fair price. Thus, they prefer to sell their produce at the village markets itself.
3. Most of the farmers in the Karbi Anglong district are very poor and thus they have no capacity to wait for better price of their produce in the absence of proper credit facilities. Farmers often have to go for even distress sale of their output to the village moneylenders-*cum*-traders at a very poor price.
4. A large number of intermediaries exist between the cultivator and the consumer. All these middlemen and *dalals* claim a good amount of margin and thus reduce the returns of the cultivators.
5. Due to the absence of market intelligence or information system in the district, farmers are not aware of the ruling prices of their produce prevailing in big markets. Thus, they have to accept any price for their produce as offered by traders or middlemen.

Suggestions for the Improvement of Road Transportation system for the Development of Agricultural Marketing in the District:

1. Considering the huge problem of congestion in the existing national highways, it is quite important that more national highways, expressways, bypasses and four-laning of congested corridors be constructed throughout the district for maintaining an improved road network.
2. The Government should design a definite financial policy for the improvement of road sector. Steps should be taken

for proper maintenance of the existing roads within the shortest possible time. As per the amendment of the National Highway Act in 1995, private sector participation in road construction project on Build—Operate—Transfer (BOT) basis should be encouraged to develop a sound road network throughout the district.

3. For developing satisfactory agricultural marketing cheaper and adequate means of transport must be developed so that farmers can take their produce in urban market or *mandis*.
4. In order to ensure a fair and satisfactory market for agricultural produce, elimination of middlemen is very much required. Such middlemen between the farmers and the ultimate consumers usually disturb the normal functioning of the market. This can be possible only when the road transportation facility develops in the district.
5. Suitable agricultural marketing structure needs an improved and adequate storage capacity in the form of modern warehouses and cold storages. Such facilities can raise the holding capacity of farmers for getting a remunerative price of their product. This can be possible only with the development of road transport system.
6. The poor bargaining capacity of the farmers arising out of poor holding capacity should be improved for getting price of their produce in the market. A good number of regulated markets should be set up throughout the district for removing the practice of exploitation of farmers by the middlemen. Weights and measures are also to be modernized.
7. Agricultural marketing co-operative society should be formed throughout the district for developing a better marketing structure.
8. Proper arrangement should be made through mass media coverage to pass correct and updated informations to the farmers about ruling prices and marketing operations.

Conclusion

Rural development is the prioritized objective of every state of India. Economic development is measured by the strength and efficiency of the linkages established in various sectors of the economy. As an agrarian economy of the district, marketing of farm produce therefore constitutes the nerves centre of rural economy. Without well-connected road facility, no economy, whether it is based on industry or agriculture can be developed. Hence, development of an area is wholly dependent on well-connected road communication facility.

References

Agriculture at a Glance, Directorate of Agriculture, Assam.
Assam Statistical Handbook, 2011.
District Census Handbook, Karbi Anglong, 2011.
Indian Economy, P.K. Dhar.

Recommendations

The ICSSR New Delhi sponsored two days national seminar on "Road Communication and its Development in North-east India and Beyond: Bridging Culture and People" was held in Department of History, Assam University Diphu Campus to discuss various issues related to road communication of North Eastern Region of India. Since there are very few researches or documentation available, the platform was very highly important one where more than 75 participants from various parts of India were present and about 40 research paper on various themes were discussed in the two days national seminar.

Some of the important outcome/recommendations of the seminar are:

1. The region required special attention from the government in terms of the development of road communication and transportation.
2. Since the region is historically and cartographically important which geopolitically shares a long frontier with China, Tibet, Bhutan, Bangladesh and Myanmar, this frontier region has experienced considerable population movement from different directions and maintained a few trade and migration routes with East and South-east Asia historically and revival of these routes is important from socio-economic point of view. Hence, the learned gathering recommended to the government to look the issue.
3. In the pre-independence period the region was highly neglected by the State except for commercial interest and in the post-colonial period was its continuation without any major changes. But the Naga and Mizo political

struggles in the 1950s and 1960s and Chinese aggression in 1962 has awakened the Indian state to emphasise significance of the roads in this region and Border Road Organisation took over a few construction of road for military interest. No socio-economic importance was received in this period.

4. The Government looks to the development of connectivity from the economic interest in 1970s onward and hence initiated the India's Look East Policy and road development initiatives started. But no interior road was developed.
5. There were some schemes of development of roads in the region. Along with this development, there were various issues—one, how the local people view the nature of the roads? Considering the density of population a functional one to facilitate movement of required goods and services or four lanes to facilitate movement of large vehicles and fast movement of personal cars? Two, how the landscape and ecology factors, as well as the volume of traffic, are taken into consideration in road widening and new constructions? Is the time a factor to explain the improvement of the communication mode? Do we have an evaluation how the time thus saved helped the economy of the region?
6. In the construction of new road and development of old road, some politics were involved. To find out the bad element of those politics many scholars suggested a transparent policy of road construction and improvement is required, which is reachable to the common people.
7. Presently some rural roads are connected by PMGSY and some roads have been constructed under MGNREGS. In these constructions, various lacuna are exposed by some scholars like lack of proper planning, lack of transparency and lack of long-term planning, which failed the objectives of these constructions.
8. In the name of widening the road and construction of new road a huge number of trees were destructed which is very highly sensible matter in the age of global warming and climate change. So proper plan in this regard is required which was a major outcome of the seminar.

9. Various rural roads, especially in hill areas are not in good shape and because of the inhabitant of these areas lagging behind socially and economically. The location like Karbi Anglong, Dima Hasao in Assam and other hill states of North-east India having these issues for which other allied issues like health care, education etc arises. The seminar recommended addressing those issues by policy-makers to erase such problems.
10. In the plain of North-east India, an important mode of transportation was neglected from the post-colonial period, the water way, which had a great importance in the pre-colonial and colonial period. Suggestions were received to revive the waterway to solve some traffic problem and minimise the transportation cost.
11. Since the North-eastern region of India geopolitically shares a long frontier with China, Tibet, Bhutan, Bangladesh and Myanmar, this frontier region has experienced considerable population movement from different directions. There was a high degree of intermingling of various streams of people, which come from East and West. Hence along with internal which connectivity, the external connectivity towards east require being re-opening for the composite development of the region, which was a major recommendation of the seminar.

Index